MOLECULES, CELLS, AND DISEASE

Springer Study Edition

JULIEN L. VAN LANCKER

MOLECULES, CELLS, AND DISEASE
An Introduction to the Biology of Disease

WITH 60 ILLUSTRATIONS

Springer-Verlag
New York Heidelberg Berlin

Julien L. Van Lancker, M.D.
Department of Pathology
School of Medicine
The Center for the Health Sciences
Los Angeles, California 90024

Library of Congress Cataloging in Publication Data

Van Lancker, Julien L   1924-
    Molecules, cells, and disease.

    "Springer study edition."
    Bibliography: p.
    Includes index.
    1. Pathology. 2. Pathology, Molecular. 3. Pathology, Cellular.
    I. Title. [DNLM: 1. Molecular biology. 2. Pathology.
    QZ4 V259ma]
RB112.V37   616.07   77-893

All rights reserved.

No part of this book may be translated or reproduced in any form without permission from Springer-Verlag.

© 1977 by Springer-Verlag New York Inc.

Printed in the United States of America.

9 8 7 6 5 4 3 2 1

ISBN 0-387-90242-2 Springer-Verlag New York

ISBN 3-540-90242-2 Springer-Verlag Berlin Heidelberg

To the memory of Monique

PREFACE

In tracing their origin and their fate, the beginning and the end of their environment, humans have often been guided by curiosity. Such concern has helped man to discover, among other things, the structure of the universe from star to atom and the evolution of life from unicellular organism to human being.

The study of disease is unique. Although it may have been inspired by the curiosity of a few, it has always been the concern of all, because preventing or curing disease has meant survival not only of individuals, but of entire nations, not only of humans, but of fellow living creatures.

If greed, force, religion, and language have been major causes of wars, diseases, more than arms, have often decided the outcome of battles and thereby have woven the pattern of history. For millennia, a large fraction of the human race believed that disease expressed the wrath of God(s) against individuals or societies. Therefore, only priests or priestesses, kings, and queens were endowed with the power of healing.

In the West, Hippocrates is credited for exorcising this concept of disease and for objectively describing and cataloguing them. The contributions of Greek physicians to Western medicine made possible more accurate diagnoses and prognoses.

Sound theories on the mode of disease production were proposed only in the latter part of the 19th century. Then, various agents causing disease, among them bacteria, viruses, parasites, and chemicals were identified. It was also recognized that, except in their simplest form, living organisms are made of cells and that the cell is the target for the injurious agent. The development of the cell theory and its extension to pathology, the science of disease, are landmarks in the history of medicine. These concepts permitted not only more accurate diagnoses, but also provided a rational basis for prevention and disease therapy. Within less than a century, dreadful epidemics (cholera, plague, typhus, yellow fever) and less widespread infectious diseases (tuberculosis, leprosy, poliomyelitis) were eradicated. During the first half of the 20th century, the role of vitamins in the diet and the diseases caused by their deficiency were identified. Scurvy, rickets, xerophthalmia, beriberi, pernicious anemia, and pellagra have disappeared wherever socioeconomic conditions do not interfere with the availability of an adequate diet or with vitamin supplementation.

In spite of progress in prevention and therapy, human beings continue to suffer and die from atherosclerosis (a cause of heart attacks and stroke), cancer, hereditary diseases, and a number of mutilating and often fatal diseases, such as multiple sclerosis and rheumatoid arthritis. In a few cases, environmental factors, such as high levels of cholesterol in atherosclerosis or smoking in cancer of the lung, have been found to contribute to the onset of the disease, but the molecules that are altered by the environmental agents and whose alterations cause the disease are unknown.

From the eve of World War II through the third quarter of this century, biologists have probed deeper into the structure of the cell. What could be seen in the light microscope only with the aid of ingenious staining techniques came in full view with the advent of transmission electron microscopy.

In this time, knowledge in molecular biology has expanded at a pace unprecedented in the history of science, except for the development of physics a few decades before and after World War I. Genetic information was found to be stored in DNA. The graceful structure of DNA was revealed by a combination of x-ray crystallography, chemical analysis, and ingenious deductions. Observations on the appearance and disappearance of enzymes in bacteria brought Jacob and Monod to propose the most daring generalization in biology: one which explained the transfer of information from DNA to messenger RNA and proteins. The Crick and Watson model for DNA and the Jacob and Monod theory form the skeleton on which all modern biology is built. What was first discovered in bacteria was shown to be true, at least in part, in multicellular organisms. Still a great deal remains to be learned.

A detailed dissection of those molecules that regulate gene expression, secure specific cellular function, and protect the cell against environmental intruders, whether they be physical (light and x-ray), chemical (toxins and carcinogens), or biological agents (microbes, viruses, and parasites), will undoubtedly result in the description of the multiple-step sequences that lead from injury to misery, from the insult to the development of a panoply of symptoms and ultimately death. If the future can be predicted on the basis of past experience, one can expect that a better understanding of disease mechanisms will result in the cure of what is now incurable.

The purpose of this book is to present the undergraduate student in the biomedical sciences with examples of modern concepts in disease mechanisms.

No attempt is made to cover all diseases and their symptoms. Inasmuch as a detailed bibliography of the subjects discussed can be found in a more comprehensive publication, "Molecular and Cellular Mechanisms in Disease." References have been omitted except for a bibliography at the end of the book.

Julien L. Van Lancker

ACKNOWLEDGMENTS

No one can trace all the sources of his knowledge or motivations. Those who have contributed directly or indirectly to this effort range from my teachers to my children. Certainly no group has been more helpful than medical and graduate students, who, by questions, criticisms, and literature searches, have often provided the material and sparked the thoughts presented in this book. I am particularly grateful to the medical students of Brown and Howard Universities, who for several years were subjected to a series of lectures in molecular pathology, and to graduate students in experimental pathology at the Universities of Wisconsin and of California at Los Angeles. The book could never have been completed without the help of Mrs. Joyce Glover who typed, Eleanor Dunham who copyedited, and Donna Maria Roberts who proofread the manuscript.

CONTENTS

PREFACE vii

ACKNOWLEDGMENTS xi

1 THE CONCEPT OF DISEASE AND ITS HISTORY — 1
Introduction — 1
Prehistoric Times — 2
Egypt, Greece, and Rome — 3
Medieval Times — 5
The Foundation of Modern Pathology — 5
    The Renaissance — 5
    The Great Schools of Medicine in the 18th and 19th Centuries — 7
Cellular Pathology — 12
The Origins of Chemical Pathology — 17
    Alchemy — 18
    Renaissance Chemistry and Medicine — 19
    Discovery of Oxygen and Respiration — 23
    The 19th Century — 28

2 DEVELOPMENT OF EXISTING KNOWLEDGE — 33
Introduction — 33
Gene Expression — 33
Cell Structure — 40
    Molecular Organization — 40
    Nucleus — 42
    Cytoplasm — 43
    Endoplasmic Reticulum — 43
    The Cytosol and Glycolysis — 49
    Mitochondria and Aerobic Pathways — 50
    Other Cytoplasmic Organelles — 51
    Cell Membrane — 51
Intercellular Integration — 53
    Microscopic Features — 53
    Nervous System — 54

	Chemical Messengers	55
	Circulatory System	56
	Conclusion	57
3	**DEFENSE MECHANISMS**	59
	Introduction	59
	First Line of Defense	60
	Blood Coagulation, Inflammation, and Immunity	61
	Blood Coagulation	64
	Inflammation	66
	Immunity	69
	Restoration of Lost Tissues	72
	Wound Healing	73
	Fractures	75
	Hypertrophy and Hyperplasia	77
4	**CAUSES OF DISEASE**	80
	Introduction	80
	Hereditary Diseases	81
	Congenital Anomalies	82
	Conflict with the Environment	84
	Trauma	84
	Toxins	85
	Biological Agents	86
	Nutritional Deficiencies	118
	Hormonal Imbalances	120
	Body Fluids and Electrolyte Imbalances	121
	Remarks on Origin of Life	121
	Water and Sodium Balance	124
	Defective Defense Mechanisms	125
	Excessive Response to Injury	127
	Conclusion	130
5	**INJURIES TO UNITS OF SPECIFICITY**	131
	Introduction	131
	Primary Injury to DNA	132
	DNA Repair	133
	Consequences of Injuries to DNA	137
	Inherited Mutations	138
	Somatic Mutations	139
	Interference with DNA Synthesis	145
	Enzymic Block	146
	Substrate Depletion	146
	Antimetabolites	146

	Injuries to Transcription	155
	Pathology of Translation	158
	Conclusion	161
6	**INJURIES TO CATALYTIC UNITS**	**167**
	About Atoms and Molecules	167
	About Chemical Reactions	173
	About Enzyme Reactions	175
	Life Cycle of Enzymes	180
	Pathology of the Catalytic Unit	182
	Absence of Enzyme Molecules	183
	Defective Enzymes	183
	Increased Enzyme Activity	184
	Defective Enzyme Regulation	184
	Defects in Coenzymes	184
	Metal Deficiency	188
	Enzyme Inhibition	189
	Disease and Substrate Alteration	190
	Substrate Deficiency	190
	Substrate Excess	194
	Conclusion	194
7	**HORMONE IMBALANCE**	**195**
	Endocrine Glands	195
	Mode of Action of Hormones	196
	Regulation of Hormone Secretion	198
	Second Messengers	200
	Hormone Degradation	201
	Hormonal Diseases	201
	Absence of Endocrine Organs	201
	Nutritional Deficiencies and Goiters	204
	Defects in Hormone Synthesis	204
	Defects in Hormone Transport	204
	Receptor Defects	205
	Regulatory Defects	205
	Glandular Hyperplasia	206
8	**PATHOLOGY OF CELL MEMBRANES**	**207**
	Introduction	207
	Function of the Plasma Membrane	208
	Transport	208
	Movement	211
	Adhesion	212
	Cell Communications	214

	Regulation of Cellular Metabolism	216
	Structure of Cell Membranes	218
	Molecular Organization of the Cell Membrane	219
	Biosynthesis of the Cell Membrane	224
	The Cytocavitary Network	226
	Endoplasmic Reticulum	227
	Lysosomes	227
	Peroxisomes	228
	Mitochondria	229
	Pathology of Cell Membranes	234
	Structural Injuries	235
	Functional Injuries	239
	Conclusion	250
9	**REFLECTIONS ON CELLULAR DEATH**	**251**
	Introduction	251
	Primary, Secondary, and Critical Injuries	252
	Programmed Death	253
	Provoked Death	255
	Molecular Mechanisms in Provoked Death	256
	The Point of No Return	256
	Catabolic Enzymes and Cellular Death	257
	Correlation of Some Morphological and Biochemical Events in Cellular Death	259
	Conclusion	262
10	**THE GREAT KILLERS, ATHEROSCLEROSIS AND CANCER**	**263**
	Introduction	263
	Atherosclerosis	264
	Pathogenesis	264
	Coronary Heart Disease	265
	Cerebral Consequences of Atherosclerosis	268
	Concluding Remarks on Atherosclerosis	272
	Cancer	273
	Clinical Pathology of Cancer	273
	Pathogenesis of Cancer	275
	Chemical Transformation in Vitro	276
	Carcinogens in Humans	277
	Cancers and Viruses in Humans	283
11	**AGING**	**288**
	Introduction	288
	Manifestations of Aging	290
	Possible Cellular Mechanisms in Aging	290

Biological Causes of Aging	**293**
Programmed Aging	**294**
Somatic Mutations and Aging	**294**
Immunologic Theory of Aging	**296**
Conclusion	**296**
Bibliography	**297**
Index	**301**

1

THE CONCEPT OF DISEASE AND ITS HISTORY

INTRODUCTION

Health is sometimes defined as a state of total physical, mental, and social well-being. In this broad sense the maintenance of health is a responsibility all individuals share.

In a more restrictive sense, health is a state of physical and mental well-being existing in the absence of incipient disease.

Disease is a departure from health which at first may not generate discomfort, but which usually leads to generalized or localized derangement, resulting in anatomical and functional distortions.

The individual manifestations of disease are often called symptoms. Although a myriad of symptoms may occur, they are often combined, thus yielding a typical pattern of distribution in the body and evolution in time. Such patterns define a disease.

The role of medicine is to maintain health by preventing and curing disease. To prevent and cure disease, one must know its causes and be aware of the sequence of steps that lead from the first injury to the full deployment of the symptoms in a typical disease pattern. In other words, one must understand disease mechanisms.

The main purpose of pathology, the science of disease, is to discover disease mechanisms. Without pathology, progress in prevention, diagnosis, and therapy is left to chance rather than to a rational attack. Yet, pathology was one of the last medical disciplines to be developed because rational concepts of disease, based on direct observation of diseased organs, have been advanced only in the last 150 years.

A brief review of the evaluation of disease concepts in history dramatizes the role of pathology in modern medicine because it reveals how ideas about pathology brought medicine from preconceived notions into a rational system of prevention, diagnosis, and therapy.

PREHISTORIC TIMES

Knowledge about the medicine of prehistoric times is scanty, and interpretation of disease concepts that prevailed then can be only conjectural. But because the biological world of prehistoric times was not essentially different from that of today, it is often believed that disease which exists today must have existed then. Although this may be generally true, environmental changes introduced by civilization have often been responsible if not for bringing about new disease, at least for modifying the incidence of some diseases.

Obviously, most of the information on prehistoric manifestations of disease stems from archaeological studies. Only those parts of the body that survive spontaneous autolysis or putrefaction can be retrieved. Therefore, much of our knowledge of prehistoric pathology results from the observations of diseased bones buried during prehistoric times and uncovered in recent centuries. Studies of prehistoric bones reveal two important aspects of disease: one relates to disease incidence, the other to the interpretation of disease mechanisms. Contrary to the popular belief that men were healthy when they lived free from the artifacts of civilization, it is clear that the full gamut of injury and reactions to injury (with few exceptions) described today in pathology textbooks must have existed in prehistoric time. Thus, excavated bones reveal signs of necrosis, inflammation, cancer, and immune reactions, and one finds indications of heart disease, arthritis, osteomyelitis, chondrosarcoma, osteosarcoma, and other diseases. For reasons unknown, prehistoric men depicted diseases in paintings, wood carvings, and clay and metal statues; men later recorded their knowledge on stone in various forms of writing. Almost every civilization recorded detailed descriptions of diseases, interpreted their mode of occurrence, and gave directions for treatment. It is believed that the first historical record of the appearance of a medicine man was found in the caves of *Les trois Frères*.

One unexpected finding among prehistoric bones is the high incidence of trepanation. It is not known why such complex, painful, and often fatal operations were performed, or even how they were done. It is believed that using stones of the appropriate sharpness and shape, one could bore small holes in the skull in 6 minutes. The primitive bone surgeon is thought to have drilled many such holes on the path of a circumference of appropriate diameter until a circular plate fell out of the skull. The incidence of recovered skulls that were subjected to trepanation is so large that it is difficult to imagine that most prehistoric trepanations were indicated. It is much more likely that the procedure reflected the belief that most diseases were caused by evil spirits which when allowed to escape would permit the body to grow healthy again.

Possibly a more accurate picture of the types of diseases that prevailed in prehistoric times is provided by observing primitive customs which by chance of history and geography escaped the benefits of civilization except for fallout. On the basis of this assumption, even for prehistoric man, the cause of disease is mystical. Men become sick because they have offended the gods, and evil spirits crowd their bodies and cause the symptoms. For example, some primitive tribes who witness the swarming worms extruding from their bodies believe that these worms cause the pain that they experience occasionally. The Indians of the Amazon advocate a blow on the head as therapy for headaches to chase the evil spirits.

EGYPT, GREECE, AND ROME

The first historical records of disease came from Egypt. A British archaeologist claims that in the early days of the Egyptian culture, bodies were often buried in the desert sand, where they became dehydrated, preserving a configuration vaguely reminiscent of the living. Bodies were occasionally brought back to the surface at the time of the floods, as if they came back from another world to visit the living.

If such events took place, they are likely to have contributed to belief in the immortality of the soul. In any event, the Egyptians developed elaborate methods of embalming and, as a result, they must have seen many gross visceral anomalies from 3000 B.C. up to the time when the soldiers of Ramses II were defeated by the Hittites. Even after the decline of the empire, the Egyptians continued for centuries to embalm their dead and to bury the bodies in painted sarcophagi.

Unfortunately, possibly because the Egyptian priests were satisfied with supernatural interpretations of the causes of disease, no attempt at theory or classification of disease seems to have been made. What is left from Egyptian medicine are mummies with anatomical anomalies, paintings of sick individuals, and papyri (Kahun, 1900 B.C.; E. H. Smith 1500 B.C.; and Ebers 1550 B.C.)—mainly case reports and formularies. They offer no systematic interpretation of the causes or mechanisms of disease production. As is the case for their contributions to many other fields of human activity, the Egyptians' approach to medical problems was practical, without attempt to mold the knowledge of the time into theories. Although medicine had been secularized in Egypt, the concept of disease remained mystical. Disease was an evil superimposed on health. Nevertheless, study of the mummies has revealed that many of the diseases prevalent today (rheumatism, atherosclerosis, appendicitis, osteosarcoma, poliomyelitis) existed then.

Early Greek medicine was not very different from the Egyptian and centered around the cult of Aesculapius. However, at the time of Hippocrates in the fourth century B.C., the concept of disease changed.

According to Sigerist, the great medical historian, little is known about Hippocrates, yet the hippocratic writings remain a fundamental document in the history of medicine.

When one attempts to summarize the contributions of the hippocratic approach to medicine, one discovers a pattern similar to that to be followed by Descartes in his *Discours de la Méthode*: the mystical concept of disease was discarded and the manifestations of disease were described on the basis of careful observation of patients, even occasionally on the basis of experimentation. All observations were recorded and organized into logical systems. The first theory on pathogenesis of disease was proposed—the famous theory of the four humors.

The complexities of the humoral theories easily equal those of modern pathogenic interpretations. In fact, the humoral theories are easier to understand if they are described in modern language.

The body elaborates four humors: blood in the heart, water in the spleen, bile in the gallbladder, and phlegm or mucus in the head. Food and drink provide the building blocks for each of these humors. The source of energy for the elaboration of these humors (which circulate throughout the body) is innate heat. In health, the humors are produced and distributed in equilibrium; in disease, the equilibrium is disturbed and too much or too little of one or more humors is produced. This interpretation of disease mechanisms prevailed through centuries with innumerable variations of the theory of the four humors.

Galen of Pergamum, a magnificent practitioner, so columniated by posterity, embodied this belief. His authority in most areas of medicine remained unchallenged through the centuries. Even Vesalius (about 1550 A.D.), the first to question some of the old beliefs, wrote about the permeability of the septum of the heart (the most important premise in Galen's theory of blood movement): "the fact that blood can pass through that apparently solid wall shows the mighty power of God."

MEDIEVAL TIMES

Except for the voluminous manuscripts composing the translation of Galen's work and some beautiful miniatures, physicians of the Middle Ages contributed little to the theories of disease. This is not surprising since dominant during these centuries was the belief in a better afterworld. Furthermore, rupture with the past resulting from invasions of Europe by barbarians had restored a mystical supernatural aura to the concept of disease.

During medieval times only 1 out of 4 newborns lived and half of the mothers died in childbirth. Epidemics could wipe out a nation's youth in less than a month, and those who survived disease contracted wounds in battle and died of infection. Only kings and saints were believed to have the power of healing. Physicians had become philosophers who gargled themselves on Galen's theories.

In the midst of the splendor of the castles and cathedrals, the savagery of the Crusades, and the mysticism of religious belief, quiet monks were recording medical documents, and mysterious men were turning witchcraft into a science. In secret alcoves alchemists were making discoveries that were to become the foundation of chemistry.

About the time Martin Luther posted his 95 theses on the door of the Schlosskirche in Wittenberg, Paracelsus had repudiated, along with the alchemist's discoveries, the existing medical theories, burned Galen's books, and attacked the academic medical faculties. Paracelsus prepared the field for a scientific medicine; but he did not propose lasting pathogenic theories of his own, and chemistry did not enter the field of medicine for many centuries.

THE FOUNDATION OF MODERN PATHOLOGY

The Renaissance

The foundations of modern pathology rest on the development of anatomy and later, its sister disciplines, histology and embryology.

Andre Van Wesele (Flemish for the Latin name Vesalius) is believed to be the natural son of Everard Van Wesele and Elisabeth Crabbe. Everard's father and grandfather were both physicians; Everard became the personal physician of the Emperor Maximilian of Austria. Everard Van Wesele had a pharmacy in a section of 16th-century Brussels that harbored prostitutes and criminals. In fact, from the window of the pharmacy one could see the Galgenberg, with its scaffold, on which criminals were hanged and left until their flesh rotted or until some investigative person removed the bodies for dissection.

This upbringing, combined with the interests instilled by his ancestors, encouraged young Andre to study anatomy. The rest of Vesalius' story was immortalized in his *Fabric of the Human Body*. One can hardly imagine how much intellectual courage was needed in those days of religious persecution to present objective observations. We gain some insight from the fact that Vesalius found it necessary to write long paragraphs denying the existence of the mystical Bone of Luz. The dogma of the time asserted that men regenerated from this bone.

The contributions of William Harvey are of as much consequence to the development of pathology as those of Vesalius. The discovery of the circulation of blood shook the foundations of the humoral theory of disease. Harvey showed that the path of blood circulation does not correspond to those of phlegm and bile circulation. A new theory—the fluid theory—came into play. Bile, phlegm, and water were in the background; blood became the central humor. It was rightly thought to be a fluid composed of several different elementary fluids, and unbalanced equilibrium of the blood was said to cause disease (dyscrasia).

Galileo modified the telescope invented by a Dutch lens maker and turned it into the compound microscope. This somewhat primitive instrument was improved by Anton van Leeuwenhoek of Leyden. Using the precious instrument he made, Anton discovered an entire microcosm of life. He saw bacteria in water, red cells traveling in the bloodstream, and he even described cross-striation of the voluntary muscular fiber.

Marcello Malpighi from Bologna put this marvelous new instrument to good use; with its aid he described important histological structures in the skin, spleen, kidney, testis, and ovary. Above all, he discovered the capillaries of the lung and thereby established the missing link in the blood circulation theory. In spite of his genius and diligence, Marcello was not treated well by his colleagues. Not only did they attack him in vituperative pamphlets, but they went so far as to sack his house, destroy his instruments, and burn his papers. One can only rejoice that today's scientific arguments are settled in a more restrained fashion.

The Great Schools of Medicine in the 18th and 19th Centuries

One would think that the science of pathology would have been developed quickly after the birth of anatomy, physiology, and histology. Many anatomical and histological descriptions of disease were published in books and atlases. But this impressive compilation of facts was not able to shake the humoral theory of disease, mainly because the linking of the various manifestations of disease into pathogenic systems was not attempted.

However, one should not underestimate the contributions to pathology of the various great schools of medicine of the 17th and 18th centuries. To name them all would be impossible, but let us consider a few.

When students were traveling to Leyden rather than Padua for their training, Giovanni Battista Morgagni, first assistant professor of theoretical medicine, later professor of pathology, published in 1761 *On the Seats and Causes of Disease*, which probably first established pathology as a science and discipline with its own immediate goals and methods.

This book contained a new concept: organs such as liver, heart, lungs, and brain were the sites of pathological manifestations of disease. Morgagni's efforts are at the origin of organ pathology. The modern pathologist, ambitious to write a textbook of pathology, can only reflect with nostalgic melancholia on Morgagni's time. Morgagni never experienced the pressures of competitive publication: his book came more than 250 years after Benivieni's book *Secret Causes of Diseases* and 30 years before the nephew of William Hunter, Mathew Baillie, wrote *The Morbid Anatomy of Some of the Most Important Parts of the Human Body*.

Dutch medicine was unknown until a tall, robust, portly man appeared whose fame was to exceed that of all physicians of his time, Boerhaave. His fame originally was derived more from his personality and his broad knowledge of sciences and languages than from genuine contribution to medicine. In fact, his contemporaries said that Boerhaave's presence was more effective than his remedies. Apparently only one experiment, a rather surprising study on the effect of heat on animals, has been credited to Boerhaave. He requested his students, among them Fahrenheit, to put a sparrow, a cat, and a dog in an oven at 73°C (in separate ovens, I presume). This study led to the remarkable observation that it took 28 minutes to kill the cat and the dog, but only 7 minutes to kill the sparrow. Although Boerhaave contributed little to medicine, he founded the School of Leyden, an offshoot of which was to become the Vienna School, which played such an important role in the development of pathology.

Boerhaave was also a chemist. He repeated some of the experiments of Robert Boyle and John Mayow without contributing anything new. He was, however, among those who recognized the attraction and affinities between various chemical elements and contributed to the enunciation of the problem. He edited numerous books, including *The Biblia Naturae,* written by his compatriot, Swammerdam.

Although Padua, Leyden, London, Vienna, and Gottingen had become great centers of medical education and of experimental medicine, Parisian doctors continued to find great comfort in their *a priori* theories or armchair interpretations of disease. French literature of the time is full of sarcastic descriptions of the doctors; the most famous caricature appears in Molière's *Le Medecin Malgré Lui.*

In the meantime, the entire political structure of Europe was remade in France, at the expense of the heads of kings, bishops, and aristocrats and the people's blood, which first ran in the gutters of Paris and in the fair fields of France and later soaked all Europe. The then existing social structure of France and Europe was wiped out.

The medical profession had its share in this purge. The existing medical corps was killed by guillotine or on the battlefield. The medical schools and medical societies were abolished. Faced with what may have been the most acute shortage of doctors in Western history, the Convention had to take steps. One stroke of the pen, after a few hours of debate, made medical history. Medicine, surgery, and obstetrics were united into a single discipline. The Convention founded three Écoles de Santé, purposely avoiding the word "medicine," and provided three hospitals for each school: one for surgery, one for internal medicine, and one for special diseases.

In these hospitals a new generation of great French doctors was trained; these men were to lay the foundations of modern scientific medicine.

Among the first of them was Xavier Bichat. One day during the rounds of the wards at the Hotel Dieu, the attendant physician of one of the patients was absent, and a young student—Bichat—presented the case in his stead.

The presentation impressed Desault (a physician of the Hotel Dieu) so much that he invited the student to live in his home and made him editor of the *Journal de Chirugie.* Later, when he had become a physician at the Hotel Dieu, Bichat spent much of his time performing autopsies. Although he died at 38, he left an impressive bibliography. The contributions of Bichat include skillful descriptions of many aspects of morbid anatomy and enunciation of the concept of tissues he called membranes.

The humoral theory of disease had lost ground since the description of organ pathology by Morgagni. It became clear that the various organs constitute sites of the manifestations of disease: for example, abscesses could be observed in different organs—lung, skin, brain, etc.—as if a unifying force existed that determined the appearance of disease at these various sites.

Many physicians of these times believed in the existence of this mysterious force. Bichat frankly rejected it and claimed that all organs were a composite of 21 tissues and that the tissues were the site of the disease. Hence, if the same tissue is present in different organs, the disease will manifest itself in identical ways in different organs.

Bichat was followed by Laënnec, Broussais, Corvisart, and many other famous physicians. Corvisart (1755–1821) was the son of a magistrate under Louis XVI. He started his career as a law student but found the profession boring. Instead of dutifully copying the writs, wills, and briefs in his father's study, he escaped the musty rooms to run to the Hotel Dieu, where he listened to Desault, who had been Bichat's teacher. Inspired, Corvisart became a physician.* He was the clinician who introduced percussion (discovered in Vienna) in France. Corvisart emphasized the importance of carefully observing disease by touching, looking, and listening. He insisted that after identification, diseases needed to be classified, and that their anatomical manifestations ought to constitute the basis of such classification. Corvisart lucidly realized that pathogenic theories could be developed only after careful observation of symptoms and clinical pathological correlation. "The physician who fails to combine the pathological physiology," said Corvisart, "with the anatomy will never be anything more than a more-or-less adroit, diligent and patient prosector. In practice he will be vascillating, unstable especially as far as the treatment of organic lesion is concerned."

Corvisart applied for a position at the Hospital founded by Madame Necker but was turned down because he refused to wear a wig. He reasoned that "respect for outward signs must not de-

---

* Corvisart was not the only one to give up law for medicine. Approximately 25 years later a rich American, Philip Ricord, born in Baltimore, went to Paris to study law, but he changed his career after listening to a lecture of the famous Dupuytren.

generate into superstition." Interestingly enough, a few years later his student, Laënnec, obtained a similar position in the same hospital, but by then the French Revolution had taken place and wigs were no longer in vogue.

Laënnec was born in Brittany in 1781. He grew to adulthood under the revolution-provisory government, the convention, the directoire, Robespierre's dictature and the Terror, the consulate, the empire, and the monarchy. One year after Waterloo, he was appointed to the Hospital Necker. A keen clinician, he constantly sought to improve methods for accurate observation of disease. He introduced direct auscultation by placing his ear on the patient's body to listen to the strange sounds generated by diseased organs. Direct auscultation was not always effective and was often uncomfortable for both physician and patient. Walking in the streets of Paris one day, Laënnec saw children playing with a wooden beam: one listening at one end, the other tapping at the opposite end. This inspired Laënnec to invent the stethoscope, which soon became the symbol of the modern physician.

A faithful student of Corvisart, Laënnec, in addition to being a keen clinician, was also versed in the anatomical manifestations of disease, as his writing shows. Although he died at 35, probably from tuberculosis, his short life was not without tribulation. The discovery of the stethoscope was hailed by some but criticized by others because it introduced mechanization in medicine and was degrading to the art.

There was a man who was to make Laënnec's life even more painful; Broussais, a huge man and a rhetorical, stout supporter of Napoleon, was already a physician at the time of the revolution. Broussais* developed his own pathogenic theory. Like Corvisart, Pinel, and Laënnec, Broussais had done many autopsies, yet he came to conclusions opposite from those reached by Corvisart and his pupils, who believed that the cause of disease varied. Since the observations of Morgagni and Bichat it had indeed become obvious that diseases were distinct entities, originating in most cases in the organ where they were seated. In Broussais' view, the primary disease was always an inflamation of the gastrointestinal canal: gastroenteritis caused by irritation. All other manifestations seen

---

* Broussais' father and grandfather had been physicians. Broussais started his medical career as a physician's aid on a pirate ship, working closely with the famous Surcouf, and part of the loot constituted his first wages. He was an uninhibited orator, sometimes called the "Mirabeau of Medicine." He shouted so loudly in the amphitheater that people heard him in the street and usually applauded. Exhalted by his success, Broussais often pursued his vituperations against his colleagues in the midst of the mob.

anywhere else in the body were secondary to this primary injury. The therapeutic implications were obvious—diet, purgation, and bleeding with leeches would cure all disease. In 1827 alone, 33,000,000 leeches were imported into France from Bohemia, which brought Broussais' opponents to say that vampires had taken control of medicine. The real tragedy is that a generation of physicians flocked to hear Broussais' flamboyant and satirical lectures rather than to listen to the soft-spoken Laënnec. In 1830, after the monarchy had been destroyed, Broussais, who hated kings, was appointed professor of general pathology at the Academy of Medicine.

In medicine, many of the teachings unfounded on observation and experience lasted for more than a millenium, whereas the most obvious facts were rejected in chorus by a well-indoctrinated faculty.

English anatomical pathology developed in a fashion very different from the development of French anatomical pathology. There were essentially two major schools—that of the Hunters and that which is sometimes called the "Guys of Guys." The Hunters were brothers. William was a polished, civilized man, a surgeon who went to Oxford to become a physician and later practiced obstetrics. In fact, William Hunter was among the first to take obstetrics out of the hands of midwives and combine the practices of medicine, surgery, and obstetrics. In spite of a heavy load in patient care, William was a keen anatomist and developed one of the first schools of medicine in London.

To William's great surprise, one evening his forgotten brother, John, came to visit him and requested an assistantship. John Hunter was somewhat of a boor with vulgar features, coarse language, and uncoordinated gestures. His lectures were desperately disorganized, yet his mind was keen and perspicacious. John Hunter is sometimes referred to as the first experimental pathologist. He studied the pathogenesis of hernia, and went so far as to contaminate himself with the agent of venereal disease by dipping a lancet in the gonorrheal discharge of a patient. As a result, he contracted both syphilis and gonorrhea. One could hardly blame him for concluding that gonorrheal discharge is responsible for the development of syphilis, an error which would be corrected only much later. John Hunter developed interesting theories on inflammation while he was a surgeon for the British Army during the Seven Years' War. He also made acute observations on gun wounds which led him to write a book on the subject. A careful anatomist and an indefatigable collector, he built one of the first museums of anatomical pathology—a contribution of considerable help to one of his students, Mathew Baillie, who published in 1794 the first atlas of pathological anatomy and morbid anatomy of some of the most important parts of the human body.

St. Bartholomew's Hospital, where Percivall Pott made his famous rounds—sometimes followed by such eminent students as the Hunters—had a considerable reputation.

Sir Thomas Guy (1644–1724) was the son of a lighterman and a coal dealer at Southmore. He printed bibles, sold them, and made a fortune by investing in the South Sea Trading Company. In 1707 he built three wards of the St. Thomas Hospital and later he erected Guy's Hospital, leaving 290,449 pounds for an endowment.

Sir Astley Paxton Cooper (1768–1841), an English surgeon, studied under Henry Cline at St. Thomas Hospital in London and attended the lectures of John Hunter. First a demonstrator in anatomy and a lecturer in surgery at St. Thomas Hospital, he moved from St. Thomas to found the medical school at Guy's Hospital and became one of its major contributors. As a surgeon he was the first to tie the abdominal aorta for aneurysm. Astley Cooper had a keen interest in anatomy and became professor of comparative anatomy at the Royal College of Surgeons. While at Guy's, Cooper inspired interest in dissection and stated that he "went to bed thoroughly dissatisfied if he had failed to dissect something during the day." Under his influence, postmortem examination became routine, and a regular staff was assigned to the service of pathology. Inspired and encouraged by Cooper, a school of remarkable young clinicians blossomed who combined daily clinical practice with the study of anatomical pathology in the dissecting room. We shall mention here only three names: Bright (1789–1858), who became famous for his studies on kidney disease and edema; Addison (1793–1860), who described pernicious anemia and adrenal insufficiency; and Thomas Hodgkin (1798–1866), who is best known for his papers "On Some Morbid Appearances of the Absorbent Glands and the Spleen," which include a description of what later became known as Hodgkin's disease.

CELLULAR PATHOLOGY

Although its foundations were laid in Italy, the Netherlands, England, and France, pathology grew into adulthood in Vienna and Berlin. Rokitansky and Virchow were primarily responsible for this. The medical world in Europe in the 18th century was not quite as closed as it is now. For example, the Vienna School of Medicine was an offshoot of the famous Dutch school. In 1745 Gerhard Van Swieten, one of Boerhaave's pupils, became the physician of Queen Maria Theresa and was soon joined by Anton De

Haen. Under their influence Austrian medicine was reorganized and medical education improved. One of the first and most famous products of the early Vienna school was Auenbrugger, who discovered the art of percussion. Unfortunately, the Vienna school did not live up to its auspicious beginning and quickly reverted to archaic methods as a result of intense dissention and intolerance in a police state.

A significant step in the history of pathology was the erection in Vienna of the Allgemeines Krankenhaus, a huge hospital that soon accommodated 14,000 patients a year. From its inception the hospital had been equipped with opulent facilities for postmortems —including a morgue, autopsy rooms, and even living quarters for a prosector, a position that had just been created. Such lavish arrangements should have augured the quick blossoming of anatomical pathology in Vienna. But the first prosector of the Allgemeines Krankenhaus, Alois Rudolf Vetter, was soon disappointed by the lack of cooperation of clinicians. He resigned, was asked to return, but died. His successor, Lorenz Biermier, was plagued with the same lack of collaboration. He was so disappointed that he began drinking and was fired. The third prosector, Johann Wagner, died soon after his appointment.

After these unsuccessful attempts to staff the autopsy service of the Allgemeines Krankenhaus, the hospital directors appointed Carl Rokitansky, a 24-year-old assistant of Wagner. Rokitansky molded anatomical pathology into a specialty. Until then, in both the French and English schools, autopsies had been performed by clinicians for the purpose of confirming diagnosis. Under Rokitansky, autopsies were centralized and the number performed was markedly increased for the purpose of collecting new information on disease. Thus, 1500–1800 autopsies were performed a year at the Allgemeines Krankenhaus. Rokitansky signed 30,000 protocols in a 40-year period. He organized his observations in a famous handbook, *Pathologische Anatomie*, which, although it was not the first textbook of anatomical pathology, was by far the most authoritative in a long time. With his book, Rokitansky provided every medical student and physician with an anatomical description of practically every disease known. Rokitansky made, however, one major error of judgment in writing his book.

Corvisart had noted that often attempts to explain the clinical manifestations of disease in terms of anatomical findings failed. He accepted these limitations and ascribed them to insufficient knowledge. Thus, rather than encouraging his disciples to build pathogenic theories, he told his students to observe symptoms and to classify disease systematically. Rokitansky observed and classified disease at a scale unprecedented, but he could not resist the temptation of proposing a broad pathogenic theory. What made this scientific mind, this keen observer, revert to the humoral theory of disease seems at first approximation incomprehensible.

Any pathologist knows that autopsy findings are often trivial and cannot explain all clinical manifestations. A case at point is diabetes. A young victim of diabetic coma may die with few or no changes detectable anatomically, even if the microscope is used. This frequent lack of correlation between symptoms and anatomical changes made Rokitansky believe that many organ changes were secondary manifestations of a humoral distortion. Thus, Rokitansky predicted that a primary injury must in all cases take place in the humors or blood which, since Harvey, had become the only respectable humor. In addition, chemistry had become a science, and Rokitansky, aware that the body's reactions were chemical, assumed that disease was caused by chemical changes in the blood. There was nothing unreasonable in such an assumption, if only Rokitansky had been able to face the ignorance of blood chemistry that prevailed at the time. Blood was said to contain fibrin and albumin, although nobody knew what these compounds were made of, and diseases were explained in terms of imbalances between fibrin and albumin. Rokitansky realized that all diseases could not be explained that way. So, in despair he postulated the existence of many other components in the blood, a proper balance of which (crasia) guaranteed health, whereas imbalance (dyscrasia) caused disease. All anatomical changes were believed to result from such blood dyscrasias.

A humoral pathogenesis of disease seemed unrealistic even in Rokitansky's days because it was already known that many of the anatomical lesions observed were—like normal organs—made of cells. It would therefore appear that Rokitansky should have logically concluded that the cell is the target of the injury and that cell loss or cell proliferation is at the origin of the anatomical distortions observed. The fact is that at that time, although Schwann (1810–1882) had concluded that all tissues were made of cells, the origin of cells through cell division had not been recognized and it was believed by Schleiden (1804–1881) and Schwann alike that cells were derived from a substance called blastema by a process akin to crystallization. It is therefore not too surprising that Rokitansky proposed that a distortion of the blastema could be at the origin of disease.

Virchow (1821–1902), a vigorous young German pathologist, helped by the studies of many others, was to correct both Schwann's and Rokitansky's theories.

Virchow was born in Schievelbein, Pomerania, and graduated

in Berlin in 1843. He lived through one of the rare liberal outbursts in Prussia and saw his country grow and later act as a threatening world power.

Virchow was a philosopher, anthropologist, and pathologist. His bibliography far exceeds any imaginable by even the most ambitious assistant professor of today. Yet in addition to these many professional activities, Virchow found the time and the courage to engage in the social struggles of his day. In 1847 he was sent to Upper Silesia to investigate the causes of an ill-defined epidemic, known at the time as Hunger typhus. When he returned, he was not content to describe the disease but vigorously attacked the social system which, in his opinion, was at the source of the problem. Virchow concluded that nothing but prosperity, education, and freedom would eradicate such epidemics; and those conditions, he added, could be attained only in an atmosphere of complete and unrestricted democracy. With these views, it is not surprising that Virchow participated in the German upheaval (as did Pasteur, in France) in 1848.

Virchow had been a student of Johannes Müller, a humanist and an inspiring teacher. Müller's students included Schwann, Henle, Remak, and Reichert. This remarkable group had had new improved microscopes for only a few years when Virchow joined them, and they were making discovery after discovery. Henle later wrote about the excitement of his days in Berlin in Müller's laboratory: "Those were the great times any day a bit of tissue shaved off with a scalpel or picked to pieces with a pair of needles might lead to important and fundamental discoveries." In Berlin, Virchow became Froriep's prosector at the Charite Hospital; he was soon promoted to chief prosector and later was made professor of pathology.

As already mentioned, in 1848 an epidemic of typhus erupted among the weavers of Upper Silesia. Virchow was sent by the Prussian government to investigate its cause. Virchow's involvement in social reforms expressed through his writings in the journal *The Medezines Reform* led to his dismissal in Berlin. He was soon offered a professorship in Wurzburg. Virchow said, "When I left Berlin in 1849 to follow the call of Wurzburg, I already brought the main ideas of cellular pathology with me, but they were not yet sufficiently clear." Once in Wurzburg, this pugnacious man devoted himself to science and accommodated himself with the ultraconservative regimes of William I, Bismarck, and William II—regimes in which he served both in the lower house in Prussia and later in the Reichstag. Virchow was granted many honors by his government. Yet one can only wonder about what went through Virchow's mind when he autopsied Friederich William of Prussia, a liberal monarch, who died of cancer of the larynx within a month after his coronation to leave the throne to the future Kaiser.

Virchow's major contribution was to extend the cellular theory to pathology. The cell theory is so fundamental to modern concepts in biology and pathology that it is worthwhile to review its history.

The cellular theory originated with Hooke, Malpighi, and Anton van Leeuwenhoek. These scientists observed that plants and tissues were divided into compartments. However, all that could be seen at that time were the cell membranes. Marked improvement in the microscope was needed before accurate recognition of intracellular structures became possible. Only about 20 years before Virchow's birth were instruments providing magnification of the order of 4000–5000 times built. Thanks to the discovery of achromatic lenses, delicate intracytoplasmic and intranuclear structures became visible.

Although Schleiden and Schwann were the first to recognize that plants and animals were made of cells, they continued to believe that cells were derived from a noncellular hypothetical substance called blastema. This theory was akin to the notion of spontaneous generation. Schwann gave detailed descriptions of the formation of the chondrocytes from this point of view:

> The formative process of the nucleus may, accordingly, be conceived to be as follows: A nucleolus is first formed; around this a stratum of substance is deposited, which is usually minutely granulous, but not as yet sharply defined on the outside. As new molecules are constantly being deposited in this stratum between those already present, and as this takes place within a precise distance of the nucleolus only, the stratum becomes defined externally, and a cell-nucleus having a more or less sharp contour is formed. The nucleus grown by a continuous deposition of new molecules between those already existing, that is, by intussusception.

Two events were to dispell the notion of blastema: the disproval of the theory of spontaneous generation and the observation of cell cleavage. In 1849 Pasteur made his memorable experiment. He placed a culture medium in two sterilized flasks, left one at sea level, and took the other high in the Jura Mountains. He opened both flasks and waited. Soon bacteria grew in the flask exposed to the polluted air of the valley, whereas the broth of the flask exposed to the pure air of the summit remained clear. After this experiment, the theory of spontaneous generation was never claimed to participate in the propagation of life.

Observation of plant cells and of the cleavage of animal eggs led Remak to recognize that all cells derive from division of parent cells. These views were shared by Nagali and many others.

Virchow extended the concept of the cellular origin of cells to all living tissues, including the diseased. To express these views he coined the aphorism, *omnia cellula e cellula.* Guided by this principle, he founded modern pathology and summarized his observations and generalizations in the book, *Cellular Pathologie* (1858).

To describe Virchow's contributions to histological pathology would be to describe most of microscopic pathology. He made many individual discoveries and also developed broad generalizations on the concepts of disease. The new pathology not only became the foundation of pathogenesis, but it also provided new and powerful diagnostic tools still used daily in surgical laboratories and autopsy rooms. To be sure, the pathology of Virchow was incomplete and often incorrect. It could hardly have been otherwise because it remained for histologists of the latter half of the 19th century and electron microscopists of the first half of the 20th century to describe the intricacies of intranuclear and intracytoplasmic structures and their alterations in disease.

It is often believed that pathology reached the summit with the work of Henle and Virchow. The endeavors of Rokitansky, Henle, Virchow, and many others constituted a monumental task and provided, for the first time, a rational approach to the mechanism of disease production—a goal so well expressed by the title of the journal founded by Henle, *Zeitschrift für rationelle Medizin.*

The anatomy and histology of disease were, however, not the only approach to pathogenesis. Other new sciences were to grow as well. These new sciences—bacteriology, experimental medicine, and immunology—which were already in the making when cellular pathology developed and grew to maturity, soon blended with the new pathology. Moreover, in the days of Virchow it became obvious that interpretation of injuries and reactions to injury could not be restricted to the cellular level but had to reach into the chemical network that composed the cells.

THE ORIGINS OF CHEMICAL PATHOLOGY

Since Virchow we have learned much about cellular constituents, their mode of biosynthesis, their interaction, and the regulation of their functions. A few diseases have already been traced to distortions in molecular composition or interaction. Most diseases will some day be describable in molecular terms. Therefore, we may well be at the threshold of a new era in pathology pregnant with information so precious that when it becomes available, it will even more than at the time of Virchow provide new clues for diagnosis and therapy. Because of the potential of the chemical approach in pathology, it might be appropriate to retrace briefly the steps of chemical pathology through history.

Alchemy

The art of chemistry developed some time after the neolithic age with the discovery of smelting, first of bronze (mostly copper and tin), later of iron and steel. Although the smelting of bronze may have evolved through the discovery that natural alloys can be melted at high temperature and cast into desired shape, at some stage of history the deliberate mixing of tin and copper in judicious proportions must have been achieved.

Although iron is abundant, it was seldom used in tool making because of its unusual properties; but once it was found that repeated heating of iron on charcoal hardens it (because of diffusion of carbon) to yield steel (iron plus 1.5–5% carbon), complex metallurgic techniques developed.

As civilization matured, new manufactured goods (jewelry, glass, dyes, incendiaries, explosives, drugs, etc.) came in demand. The production of these materials was extremely lucrative. Therefore, to protect their trade, manufacturers described the steps involved in code language, "formulas" not to be understood by others. Although the empirical approach laid the foundation of chemistry, the lack of an experimental approach prevented these artisians from discovering the laws that govern chemistry. Instead, they interpreted their findings in terms of prevailing concepts of the universe. Philosophers of those days listed seven planets (the sun, Venus, the moon, Mars, Jupiter, Mercury, and Saturn); seven metals (gold, copper, silver, iron, quick silver, tin, and lead); and seven days. They believed that in some ways planets, metals, and division of time were all interrelated and chemistry became entangled with astrology. The "world soul" was believed to be made of spirits dwelling in everything, and substances extracted with fire, or other means, were thought to be manifestations of the spirits.

Since metals came from the bowels of the earth, it was then logical to conclude that it should be possible to extract the "fecundating substance" from metals provided, of course, that the search be done under the appropriate planetary interrelation.

Inspired by this philosophy, the alchemists departed from their original pragmatic approach and used their skill to uncover the elixir of life and a method of transmutation of metals. Needless to add that even the most imaginative combination of interaction of planets, metals, and days, the most ingenious extraction procedures, or the most exotic mixture of substances never yielded the philosopher's stone nor the elixir of life. However, during their search, alchemists uncovered numerous procedures and many chemicals that would later become the foundation of the science of chemistry.

In fact, a proverb of the Middle Ages stated that behind every alchemist there is either a physician or a soap maker, illustrating the need for most alchemists to provide their clients with consumer goods more mundane than the elixir of life, the philosopher's stone, or potable gold.

Even in medieval times, some alchemists recognized the limitation of their approach. For example, Albert the Great contributed new techniques and new chemicals and admitted that the procedures used by the alchemists could at best give a "special stain" to the metals treated rather than transmute them.

Renaissance Chemistry and Medicine

In 1493, just 1 year after Columbus discovered America, Wilhelm Bombast von Hohenheim, a physician, had a son whom he named Philipp Theophrastus after a disciple of Aristotle. Hohenheim's wife died shortly thereafter, and his son was his only companion. Together they walked the mountain trails of Switzerland to see patients, and as a result young Theophrastus was plunged into the practice of medicine at a tender age. For a while von Hohenheim settled in a small mining town equipped with smelters. It may have been there that Theophrastus became interested in chemistry. When of age, Theophrastus went to Ferrara where he formally learned medicine. One of his professors was Leoniceno, who decennia before Frascatorius had described venereal pestilence. Once he had obtained his degree, according to the custom of the time, von Hohenheim changed his name and chose to be called Paracelsus.

In those days, medicine was learned and taught from reading books—those of Aristotle, Hippocrates, Galen, and others. Experimentation was nonexistent, observations were subservient to accepted theories.

But those were also the days when Martin Luther pinned his theses on church doors, spoke vitriolic words against the sellers of indulgences, translated the Bible in German, renounced his vows of chastity, and rejected the authority of the Pope. This challenge to established values stirred a spirit of rebellion in the minds of many in Germany and elsewhere. That Paracelsus was carried on the crest of these waves of rebellion is not unlikely because he was, like Luther, endowed with a pugnacious personality and brought up in the midst of the plebes.

However, Paracelsus' attack on existing medicine was not all rebellion, it included genuine concern for truth and innovation, a concern Paracelsus shared with some of the most illustrious of his contemporaries. Indeed, in those years Guttenberg invented printing, Columbus returned from America, Copernicus reshaped our understanding of the universe, and Vesalius dissected bodies at the risk of death at the hands of the Inquisition.

Whatever the cause of his motivation, it is certain that Paracelsus rejected books as a source of knowledge and tried to learn about disease on his own by direct observation. Before he settled to practice in Strasbourg and Basel, he traveled around the world healing the sick when he could. He said, "If a man wishes to become acquainted with many diseases, he must seek forth on his travels. If he travels far, he will gather much experience and will win much knowledge."

When he had finally grown tired of traveling, Paracelsus set up practice in Strasbourg. Soon his reputation spread to Basel where an influential printer, Frobenius, had a leg injury that became gangrenous and for which the faculty had recommended amputation. Paracelsus was consulted, and thanks to his skillful therapy the leg was saved. One of the guests of Frobenius was Erasmus, who suffered from gout and kidney stones; he, too, consulted Paracelsus and was cured by him, or at least he thought so. Erasmus wrote to Paracelsus, "I cannot offer thee a fee equal to thy art and learning. Thou hast recalled from the shades, Frobenius, which is my other half. If thou restoreth me also, thou restoreth each through the other. May fortune favor that thou remain in Basel."

Having had such illustrious patients, it was not surprising that the fame of Paracelsus spread all over the city and that as soon as the position of town physician became available he was offered the job. Paracelsus accepted with delight because the position included a professorship in the medical school. Instead of using the graceful, precise, concise, but out-of-touch Latin language, Paracelsus chose to teach in German. At the time this alone was an outrage. But not content with this first attack on established values, Paracelsus further upset the respectable, conservative faculty by preaching a new medicine of his own rather than the teachings of the ancients. Paracelsus was scorned, persecuted, and never succeeded in building a school of thought. It is said that the character of Doctor Faust in Goethe's drama is at least in part inspired by Paracelsus. Indefatigable while alive, Paracelsus did not live to a ripe old age. He died in Salzburg Hospital (1564) before he was 50, exhausted by his struggles.

Anyone who, like Paracelsus, referred to his colleagues as a "misbegotten crew of approved asses" could hardly expect their enthusiastic support. Neither was he without bringing on himself some of the trouble that befell him. In a city in which the entire medical profession was after his skin, after losing a suit (a matter of 94 gulden), Paracelsus rebuked the courts in such vigorous language that his life was soon in danger and he had to flee. Still, he was a man with a vision, much ahead of his time.* For example, he said,

> My travels have developed me; no man becomes a master at home, nor finds his teacher behind the stove. Sicknesses wander here and there the whole length of the world. If a man wishes to understand them, he must wander too. A doctor must be an alchemist, he must see mother earth where the minerals grow. And as the mountains will not come to him, he must go to the mountains. It is indeed true that those who do not roam have greater possessions than those who do; those who sit behind the stove eat partridge, and those who follow after knowledge eat milkbroth. He who will serve the belly—he will not follow after me.

Paracelsus believed that changes in the body were chemical and that illness must be treated with chemicals.

Like Ponce de Leon, Paracelsus looked for the elixir of life, and also like Ponce de Leon he didn't find it. But for neither was this search in vain: the Spaniard discovered Florida; Paracelsus introduced laudanum, salts of mercury, lead, arsenic, and sulfur in the therapy for disease, and many of these substances are still used in therapy throughout the world.

Paracelsus' endeavors were not completely wasted. Indeed, because of him superstition in medicine received an irrevocable blow: it became difficult to convince victims of gout that they would be cured by the sound of a flute, victims of scrofula no longer believed that the touch of kings could heal their ailment, and if old men continued to believe that they could be restored to youth by young girls exhaling their warm breath on their bodies, others would not grant their hopes much chance of coming true.

---

* Paracelsus' foresight has been best expressed by Browning, "But after they will know me, if I stoop into a dark tremendous sea of clouds it is but for a time, I press God's lamp close to my breast, its splendours soon, or late will pierce the gloom. I shall emerge one day, you understand me, I have said enough."

Alchemy was to survive as a mysterious combination of astrology and technology long after Paracelsus. Rudolph the II of Germany (1576–1612) devoted most of his fortune to a search for potable gold. He surrounded himself with an academy of physicians that included Libovius. Although he may have read papers on the *Aurus Potabile*, Libovius also wrote a valuable textbook, *Alchymia*, which described known procedures and chemicals.

In the Renaissance when chemistry became a science, some physical chemists of great reputation appeared on the scene, including Van Helmont, Boyle, and others.

The German Rudolph Glauber deserves special mention because he discovered sodium sulfate, arsenic, and zinc chloride. He was the first to distill ammonia and hydrochloric and sulfuric acids, and he wrote an encyclopedia of chemistry. Yet he was called an "errant knave" by Oliver Cromwell because he persisted in the alchemist tradition to sell "secret formulae" and apparently sold the same one to different manufacturers.

Although the remedies discovered by Paracelsus, Van Helmont, and others were useful at the time, chemistry did not permeate medicine for at least another 250 years. Priestley, Cavendish, Lavoisier, Dalton, Berzelius, Werner, Mendeléeff, and Arrhenius had to lay the foundations of inorganic and organic chemistry before the chemistry of the living could be investigated by Claude Bernard, Liebig, Fischer, and many others.

Van Helmont, a Capuchin friar who became a physician, could never completely dissociate religious mysticism from scientific observation in his interpretation of disease. Like Paracelsus, Van Helmont believed that all physiological processes were essentially chemical. But in his mind the chemical process was in some way activated by a special ferment called gas, itself controlled by a superior spirit called blass. All the blasses of the body were activated by a motivating force called the *animae sensitivae motivae que*, known to be located in the pit of the stomach because a blow in that region destroys consciousness.

In spite of this mystical interpretation of medicine, Van Helmont made great contributions. He discovered carbonic acid, which he called the *gaz sylvestre*, introduced the gravimetric analysis of urine, studied bile and gastric juice, and even predicted some of the modern notions of humoral immunity.

The Leyden school added order to Van Helmont's interpretation of physiological chemistry. Sylvius and his pupils Willis, DeGraaf, and Swammerdam were responsible for this improvement. Sylvius recognized the importance of saliva and pancreatic juice in digestion and was one of the first to advocate the identity of organic and inorganic processes, a concept which would be fully accepted only when Wöhler synthesized urea. Thomas Willis, the son of a Wilshire farmer, made the first accurate qualitative examination of urine and recognized the characteristic sweet taste of diabetic urine. De Graaf studied pancreatic juice by making temporary pancreatic fistulas and he studied bile by making artificial biliary fistulas.

Discovery of Oxygen and Respiration

Galen proposed that respiration served to cool the heat generated by the heart, which without respiration would burn itself into smoke. Respiration was considered to be a means not of combustion but of refrigeration. Harvey's discoveries of blood circulation established that dark blue venous blood is changed into red arterial blood in the lung. Rejection of Galen's view and acceptance of Harvey's findings were at the origin of the search for the agent that changes venous into arterial blood. Boyle had shown that air was needed for combustion and for life. Experiments of Vesalius and Robert Hooke in which a blow pipe was attached to the trachea of a dog after opening the thoracic cavity had shown that insuflation of air could keep the animal alive. In about 1669 Loewe injected dark venous blood into the arteries of an inflated lung, observed that it turned red as it passed through the lung, and concluded that the blood binds to air present in the lung. Only 100 years later was oxygen discovered by Scheele in Sweden and Priestley in England, and then Lavoisier clarified the role of oxygen in respiration.

Clearly, a knowledge of the composition of the atmosphere was indispensable to an understanding of the body's metabolism. Although it had long been suspected that inhaled air was responsible for the red color of arterial blood, air was believed to be made of a pure gas; many believed the purpose of breathing was for cooling the blood, after the teachings of Galen and Aristotle. Air was not suspected to be in any way responsible for burning. Inflammable materials were said to contain a special substance phlogiston, which under appropriate circumstances was released into flames causing combustion of the material that harbored it.

Rutherford, a physician and a professor of botany at the famous school of Edinburgh, discovered that air was made of two gases. Thus, when he circulated air over lime, a portion of the gas was retained and reacted with the lime to give it a milky appearance. Later it was found that a gas that reacts with lime is CO_2. If a small animal was left to breathe in a jar containing air and after its death all the CO_2 was removed, Rutherford found that a large amount of gas (approximately two-thirds) was still present in the jar. That gas was later found to be associated with nitre and was therefore called nitrogen. Oxygen, the gas indispensable for animal life, had not been discovered.

An English clergyman was to make this important finding, which changed the trend of chemistry and biology for all times to come. Yet it would seem that the properties of oxygen could have been known long before. Indeed, alchemists and later chemists had known that mercury turns red when heated in presence of air. The process was referred to as calcination. The red mercury released gas when heated again, but none of the early experimentors associated the released gas with air itself.

Priestley was a minister of what was then a new and controversial church: the Unitarian Church. He taught school and ran a parish before he became famous for his scientific accomplishments. At one point of his career the clergyman settled in a village close to Leeds near the brew house of Gakes and Nell. Such a curious man quite naturally went to watch the huge vats in which the barley was left to ferment, yielding gas bubbles in the process. Since it was impossible to collect the emanating gases on the premises, Priestley prepared the gas at home and mixed the gas with water. He soon reported his findings to the Royal Society, whose members all acquired a taste for this remarkable new artificial beverage, which was as pleasing as some of the water collected at natural springs.*
In the process, Priestley had devised a simple but effective method of collecting gases. A bottle was placed over a jar containing mercury, and the gas generated in the flask was trapped in the bottle. Priestley developed lenses that were placed between the sun and the material to be ignited, focusing the rays on the target. With this system, he heated red mercury and generated gas as many had done before him, and by chance (as Priestley himself admitted) he decided to test the effect of this new gas on the flame of his candle. To his surprise, the flame grew brighter.

It had been known for some time that when mice were placed in a jar in which a candle was kept burning, they soon died. Conversely, the candle would not burn in a jar in which mice had been breathing. Therefore, it was quite logical for Priestley to plan experiments to find out what effect the new gas would have on the life span of mice placed in a jar. In those days, one could not buy

---

* In those days it was believed that mineral water could cure scurvy. So when Lord Sandwich (then the first Lord of the Admiralty) learned of Priestley's discovery, he hastily built two shops to provide the Navy with the new miracle cure. The Royal Society, thoroughly impressed, gave Priestley the highest honor they could bestow, the Copley medal.

mice to perform experiments, and Priestley had to set traps to catch his own mice without hurting them. He placed one mouse in a jar containing air, another in a jar filled with the new gas. It has been said that the clergyman watched the two jars while playing the flute. The mouse in the jar containing the new gas lived longer than the one in the jar containing ordinary air. Priestley had discovered the cause of combustion and the gas responsible for animal breathing. He then tested the new gas on himself and found that he breathed easily and became slightly euphoric.

Although Priestley had not demonstrated the combination between metal and oxygen in the process of combustion and he did not relinquish the theory of phlogiston, he did suggest the therapeutic uses of oxygen and foresaw the use of the gas in the blowtorch. While Priestley was making his landmark observations, the American War of Independence had started and the French Revolution was soon to follow. Priestley became an enthusiastic supporter of the American and French revolutions and vigorously refuted Burke's evaluation of the tragic events that shook France in those days. While Priestley was celebrating Bastille Day in 1791 in his home in Birmingham, a mob appeared and looted and burned the premises. His radical views alienated all his supporters. After spending 2 years in London where he felt ostracized, he decided to seek refuge in the United States.

When he arrived in the United States, Priestley was received as a hero. After settling in Philadelphia he preached to a congregation that was later to include John Adams and Thomas Jefferson.

Until Lavoisier's time, chemistry was a rather mystical, mysterious science. All matter was thought to be composed of a fundamental substance that could be modified by elements, spirits, and bodies; and phlogiston was the soul of all matter. The elements included earth, fire, air, and water. Three different types of earth were distinguished: mercurial, vitreous, and combustible. The four spirits were sulfur, mercury, arsenic, and ammoniac. The six bodies were gold, silver, copper, lead, tin, and iron.

Lavoisier made two major contributions to chemistry. He simplified the nomenclature, making chemistry accessible to anyone who could read, and he provided the factual evidence needed to bury the concept of phlogiston.

Early in his career, Lavoisier had doubts about the phlogiston theory and preferred instead the "caloric" or "heat" theory, which proposed that air contained a "fluid" that permeated the pores of all inflammable matter. The theory of the caloric could hardly be considered a great step forward because it only replaced one unproven and untestable theory with another unproven and untestable theory. Nevertheless, Lavoisier differed from his fellow scientists in that he saw the necessity of testing the theory of combustion. Although it seems certain that as his thoughts on combustion were maturing, Lavoisier must have become convinced of the need for carefully measuring the elements before and after combustion, it would appear that Lavoisier had no precise plan for attacking the problem until after he met Priestley.

After having been a teacher, a minister, and an amateur scientist, Priestley was allowed to spend all his time performing chemical experiments thanks to the philanthropy of an English Lord who shared Priestley's political views. In 1772 Priestley discovered oxygen, and in 1774 the Lord took Priestley on a trip to the continent. Priestley met Lavoisier in Paris. He told Lavoisier about his experiments with the red mercury and how heating the powder had produced gas that brightened the flame of his candle, kept mice alive, and gave him a feeling of well-being. Surely the conversation that took place in Lavoisier's Paris hotel on the Rue de La Madeleine must have been exciting and memorable. Yet when later Lavoisier "rediscovered" oxygen he gave no credit to Priestley—an omission that posterity has had a hard time understanding.

As soon as Priestley had left, Lavoisier went to work, equipped with what were then among the most accurate balances available. He borrowed them from the French Mint, and some of the most famous physicists had helped to build them. This careful chemist, inspired by Priestley's findings, started the *experience des douze jours*, or the "Lavoisier experiment." He heated mercury in a closed system, weighed the air that was left in the system, and after calcination weighed the red mercury he had obtained and proved that the mercury had gained in weight exactly what had been lost from the atmosphere in the closed system. He recognized that the remaining gas was nitrogen and further demonstrated that the portions of the atmosphere bound to the mercury could be released and that the released gas had the properties already described by Priestley. Lavoisier, who like many Frenchmen was prone to invent new words, called the gas "oxygen," or acid former. He concluded that combustion was due to the combination of the metal with the oxygen.

These views were not to be accepted at once. One previous observation credited to Emperor Francis I of Austria was used against Lavoisier's concept. In an experiment which only Emperors can afford and which reveals the vanity of our riches, Francis I had observed that heated diamonds volatilize, leaving no trace. When Lavoisier repeated the experiment in a closed system, heating the diamonds with sun rays in the fashion of Priestley, he showed that they were in fact converted to CO_2.

Lavoisier published his famous *Traité Elementaire* in 1789, the year when Marianne after centuries of painful labor delivered a bloody and hateful revolution that killed her own children. In 1794 Lavoisier was arrested by the agents of the *Terreur*. Marat, an unsuccessful chemist who had not been made a member of academy because of Lavoisier's objection, was among his public accusers. Lavoisier died in 1794, the same year that Priestley landed in America. But before he died he had extended, in collaboration with Laplace, his research on the calcination of metals and combustion into studies of respiration.

If it is true that Laplace is the French Newton, it is also true that, like Tallyrand, he managed to survive under numerous diversified and sometime extreme regimes: Louis XVI, the Convention, the *Directoire*, the *Terreur*, Napoleon, and the Restoration. Laplace is famous more for his achievements in mathematics and astronomy than for his studies in physiology. Nevertheless, he built a calorimeter, a modification of the one described by Black; and in one of the greatest collaborations in science, the Laplace calorimeter and the Lavoisier gasometer were used simultaneously to study oxygen intake and CO_2 production. In these experiments Lavoisier demonstrated that respiration is simply a special form of combustion, and he wrote, "on the one hand that an eminently breathable air (oxygen) is absorbed and on the other the lung puts out in its place a portion of a gaseous form of (carbonic acid) equal in volume."

Guided by these findings, Lavoisier soon concluded that in animals heat results from combustion produced by respiration, that respiration is a form of combustion occurring in the lung, and that the heat that results from it is communicated to the blood and from there to the entire animal system. He compared the animal body to a machine with three regulatory principles: respiration, which consumes oxygen and carbon and produces heat; transpiration, which increases or decreases depending on the need for reducing the heat produced; and digestion, which restores to the blood what it lost in respiration and transpiration.

If it is true that captured by the needs and demands of his time, Lavoisier helped to collect taxes as a member of the *Ferme Generale* and helped France to produce ammunition, it is also true that he had a profound understanding of the needs of mankind.

In conclusion, the experiments of Lavoisier and Laplace demonstrated that respiration is a form of combustion, the products being carbon dioxide and water. We have seen that Lavoisier believed that the entire process of combustion took place in the lung. It was up to LaGrange to demonstrate that the dissolved oxygen of the inspired air picks up the carbon and hydrogen atoms from tissues after the blood runs through them.

The 19th Century

In the 19th century at least three men were to play key roles in the development of the chemistry of disease: Claude Bernard, Louis Pasteur, and Paul Erhlich.

Claude Bernard was the son of a Burgundian winemaker. After receiving a classic education in French grammar and high schools, he tried his luck at playwriting. Encouraged by the reception of his first comedy in a small theater of a small town, he took his chances in Paris, only to have the critics suggest that he go to medical school instead of writing plays. That he did. He studied at the Hotel Dieu and the College of France, where he was one of Magendie's pupils.

Francois Magendie had succeeded Laënnec in the Chair of Medicine at the College de France. He was among the first to depart from the vitalistic theories in physiology. At that time physiology was dominated by the notion that a special "vital force" was responsible for the phenomenon of life. Even Bichat had yielded to that concept, although he modified it by proposing that the vital force was the sum of various forces distributed in different organs where they are responsible for the various vital functions. Bichat further clearly expressed the notion that life was a conflict between vital and physical inert forces, vital forces guaranteeing life, physical forces leading to death. The inescapable corollary of such a philosophy was that the laws regulating biological functions had to be different from those regulating reactions in the inorganic world. Magendie advocated that this was not so and proposed for the first time that the understanding of the mechanism of life could best be reached by systematic experiments. Magendie's experiments in physiology clearly proved his point: namely, that the laws which govern the inorganic universe also obtain in the universe of the living.

Claude Bernard extended Magendie's approach to the study of physiology to that of disease, and as a result he was to physiology and physiopathology what Pasteur was to microbiology and biochemistry. Claude Bernard was born in 1813 and lived until he was 65. When Magendie retired, Bernard took over his chair at the College de France. His contributions are too numerous to be described here. They include investigation on the role of the liver in sugar metabolism and the effects of carbon monoxide and carbon dioxide in asphyxia.* In addition to laying the foundation of physiology and physiopathology, Bernard provided biologists for all generations to come with a method for investigating life which he described in his famous book *Introduction a la Medicine Experimentale*.

Pasteur, also a Burgundian, was a chemist who had a knack for approaching practical industrial and medical problems, solving them, and providing remedies. Thus, he studied diseases of wine, beer, silkworm, sheep, chicken, and man (including anthrax, rabies, and many other infectious diseases). He laid the foundation of stereochemistry with his investigations on tartaric acid, enzymology with his studies on fermentation, and bacteriology and immunology with his studies of infectious disease in animals and humans.

It had been known for some time that the chemical compositions of tartaric and racemic acids were identical. Yet when crystals of tartaric acid and racemic acid were exposed to polarized light, different results were obtained. With tartaric acid, the beam was deviated to the right; in contrast, racemic acid had no effect on the beam. By developing ingenious methods of purification, Pasteur succeeded in separating two kinds of tartaric acid from the racemic acid: one deviating polarized light to the right, the other deviating it to the left. Pasteur proposed that the properties of deviating light were related to the structure of the crystals and that the atomic arrangements of both forms of tartaric acid were identical in every way except that one form was a mirror image of the other.

Now that the wines of France have conquered the markets of the world, it is difficult to imagine the tribulations of winemakers before Pasteur completed his studies on fermentation. Until Pasteur discovered the causes of disease in wine, one could never be sure whether the final product would taste like sweet wine or vinegar, appear clear or cloudy. The principal reaction in the formation of wine from grape juice is the conversion of glucose and fructose to ethyl alcohol and carbon dioxide. Usually small amounts of glycerol, acetic acid, succinic acid, and alcohols are also formed. Conversion

---

* All of Bernard's work was done in dire conditions,—in poorly lit, dusty, damp rooms with homemade equipment and low budget. An anecdote recounted by Sigerist illustrates the miserable condition under which Bernard had to work. Apparently, one morning a cannulated dog escaped from the laboratory. The sight of the distressed animal scandalized the inhabitants of the neighborhood, who called the police only to discover that the dog had, in fact, been stolen from the police chief. It speaks well for the representatives of the law of those days in that the robbed police chief invited Bernard to work in his precinct under his protection.

of sugar to alcohol is not spontaneous but is catalyzed by the wine yeast *Saccharomyces elipsoideus*, which is present at the surface of the grapes, with, of course, many other kinds of yeasts.*

At the time of Pasteur it was known that fermentation resulted from the conversion of sugar to alcohol and that yeasts were involved in the conversion. The role of yeasts was, however, not clear; for example, Liebig claimed that fermentation resulted from the decomposition of the yeast. Pasteur started his investigation on fermentation with a prejudice, which in some ways was derived from his studies in stereochemistry. He had observed that the amyl alcohol produced in some forms of yeast fermentation had optical properties, whereas amyl alcohol produced synthetically did not. Similarly, when racemic acid was exposed to bacteria of the penicillin family, the solution which at first had no effect on polarized light developed optical properties. Such results indicated that one of the stereoisomeric forms of tartaric acid was used by the bacteria but the other was not. This rigid specificity in the usage of nutrients suggested to Pasteur that the yeasts involved in fermentation were alive because he felt that only life could impart such specificity. Guided by this assumption, Pasteur attempted to separate various kinds of yeasts, each with special fermentation properties, and succeeded in culturing yeast in a solution containing mainly sugar and salt.†

Pasteur further demonstrated that many of the diseases of wine resulted from parallel parasitic fermentations catalyzed by agents other than the wine. He isolated an agent, to be called a microbe, that catalyzed the fermentation of sugar to lactic and butyric acids. He demonstrated that the production of vinegar from wine results from the presence of a microorganism called *Mycoderma aceti*. He established that the parasites could be killed by heating, a process known now as pasteurization, which prevents wine diseases.

---

* In white wine the juices are separated from the grapes before fermentation begins. In red wine fermentation occurs partly in presence of the grapes for the purpose of extracting tannins and pigments. In dry wines very little sugar (15%) is left after fermentation. More sugar is left in sweet wines. The sugar content depends on the amount of sugar present initially and the type of yeast used in fermentation. In general, yeasts are killed when the alcohol concentration reaches 15%, although some can tolerate higher alcohol concentrations.

† Of course, such findings negated the fermentation theory of Liebig. To convince Liebig, Pasteur offered to show him 1 kg of yeast cultured *in vitro*, and the French Academy offered to defray the travel expenses of Liebig so he could, with his own eyes, witness the marvels of microscopic life. More confident in the power of his brain than in visual evidence, Liebig refused to go to the city which was then well on its way to becoming *la ville luminere*.

Even when it was generally accepted that microorganisms can cause disease, the origin of these microorganisms remained in question. Pasteur believed that the yeast was present in the air and deposited on the grape's skins; others claimed that they were derived through spontaneous generation. We have already discussed how Pasteur resolved the argument.

The concept of infectious diseases of wine was extended to beer, to silkworms, and finally to humans.* Because of it, new sciences (biochemistry, bacteriology, and immunology) developed, and new methods (aseptia, antiseptia, chemotherapy, pasteurization, immunization, etc.) were applied saving countless numbers of lives all over the globe generation after generation.

Although Virchow foresaw the importance of chemistry in the understanding of disease mechanism, Ehrlich laid the foundation of modern chemical pathology. There is no man who is more closely connected with the development of biochemical pathology than Paul Ehrlich (1854–1915). Much of Ehrlich's life was spent in Virchow's Hospital La Charite. Ehrlich was an assistant of Friedrich Theodor von Frerichs, who had replaced Schönlein at the Charite Hospital in Berlin in 1815. Frerichs was one of the first people to study fatty liver and demonstrate its relationship to diet deficiency. He used what were then modern chemical methods to investigate his patients and demonstrated, for example, the release of leucine and tyrosine in urine in cases with acute yellow atrophy. He analyzed gallstones and classified them according to chemical composition. At the Frerichs' school Joseph von Merring and Oscar Minkowsky produced diabetes experimentally for the first time. Needless to say, this aggressive interest in a field overlapping with that of Virchow did not inspire sympathy on the part of the latter; in fact, Virchow and Traube developed a rather lasting dislike for Frerichs, a dislike which was passed on to his assistant Ehrlich.

Ehrlich made great contributions to chemistry and medicine alike. For example, he applied the Kekulé formula to the benzene ring, and this concept is still at the base of much of our understanding of organic chemistry. Familiar as he was with the nature of chemical reactions, it was not surprising that Ehrlich was the first to claim that specific substances in the living organism combine with specific chemicals introduced in it. This was, in fact, suggested by the selective staining properties of certain portions of the cell.

---

* One of Pasteur's disciples, Koch, was the first to demonstrate beyond any shadow of a doubt that bacteria cause disease. Koch also proposed a set of rigid rules for establishing the bacterial cause in infectious disease, and among many other discoveries, he identified the agent of cholera and tuberculosis.

Ehrlich exploited this notion more than anybody, and when analine dyes became available he studied their properties and developed innumerable new staining techniques for tissues. This remarkable contribution greatly facilitated histological diagnosis. However, Ehrlich's interests extended far beyond histological techniques: the germ theory was in his day firmly established and Ehrlich dreamed of synthesizing chemicals that would attack microorganisms specifically without damaging normal cells. He chose to attack the spirochete of syphilis, and after preparing 605 ineffective compounds, he finally made a 606th compound that for a long time was the only chemotherapeutic agent effective against this spirochete. Modern chemotherapy is based on those experiments. Ehrlich also showed that bacteria secrete toxins which elicit the formation of antitoxins in the body. Ehrlich was the first to investigate the chemical nature of these antitoxins and thereby laid the foundation of the sciences of immunology and serology.

DEVELOPMENT OF EXISTING KNOWLEDGE

INTRODUCTION

Two thousand years elapsed before the humoral theory of Hippocrates was replaced by the cellular theory of Virchow. The progress made in medicine in the last 100 years has not displaced Virchow's theory but has seated it on a solid foundation of facts and expanded it. Because we now know what cells look like and the chemicals they are made of, we can identify, at least in some cases, the targets of injury and reconstruct the sequence of steps that result in those cellular changes that cause disease.

GENE EXPRESSION

We shall be concerned only with mammalian cells and for the moment ignore unicellular or other types of living beings. All cells have a specific function: brain cells store information, muscle cells contract, cells of the hypophysis secrete hormones, lymphocytes make antibodies, and liver cells perform a multitude of chemical conversions. Although the frequency and the circumstances in which cells divide may vary from one tissue to another, most cells are also capable of reproduction.

The study of cellular chemistry has revealed that the cell is composed of an integrated population of molecules of various sizes, and the morphology and the function of the cell and its organelles are determined by the interaction between specific molecules. Thus, cellular differentiation and cell division result from the programmed interaction of a multitude of molecules of various sizes.

The molecular composition of each cell determines its structure and function. Thus, the human organism is not composed of single identical populations of cells, but of many different cells with highly specialized functions. These functions are integrated to secure optimal performance as we know it. Under optimal conditions, in humans such performance includes development into a mature being and maintenance of that status in the face of environmental threats.

The differences in cell structure and function result from differences in the molecular mosaic that composes the cell. Consequently, in most if not all cases, the molecule is the unit target of disease.

Although variation in the molecular population of a given cell

FIGURE 2-1. Steps in meiosis.

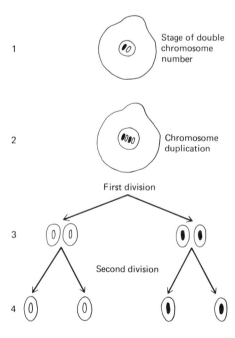

is minimal, the cell is not static. Most molecules are either partially or totally degraded and need to be replaced. Degradation takes place because the molecules are consumed during function or altered by the environment, or because their degradation is programmed. The purpose of this book is to describe how injuries to molecules affect cells, organs, and ultimately the entire organism.

All body cells are derived from one spermatocyte and one ovum. Although every cell of the human body has 46 chromosomes, each of the two germ cells has only 23. This reduction in the number of chromosomes occurs in a process called meiosis (Fig. 2-1). After fecundation the spermatocyte penetrates the oocyte (Fig. 2-2) in a process akin to phagocytosis, and the fecundated cell contains 46 chromosomes; 44 are autosomal and 2 are sex chromosomes. In the female each sex chromosome has the shape of an X, whereas in the male 1 sex chromosome is of the X type, the other is a Y type (Fig. 2-3). In the female, one of the X chromosomes is repressed and appears on stained sections of blood cells as a small mass of heterochromatin called a drumstick (Fig. 2-4).

As the human embryo grows, each new cell contains the 44 autosomal and the 2 sex chromosomes. Consequently, every cell of the body contains the same DNA. At first approximation, one would therefore anticipate that every cell would have the same protein mosaic. This is obviously not the case. Liver cells are very different from kidney cells; skin cells are different from cells of the intestinal mucosa. Thus, as the embryo develops from the fecundated ovum,

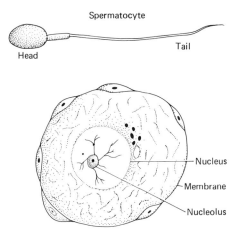

FIGURE 2-2. Schematic representation of oocyte and spermatocyte.

at various stages of fetal development some cells express a portion of their DNA and suppress others in an orderly fashion in time and space. For example, the gene for expression of hemoglobin is expressed only in the bone marrow cells destined to become red cells and is suppressed in all other cells of the fetus and the body (Fig. 2-5). The distinction between expressed and "repressed" genes gives each cell type its unique protein mosaic. Therefore, this selective process in gene expression is referred to as "differentiation." The understanding of the mechanism that turns genes off or on is central to an understanding of life of unicellular and multicellular organisms.

Our knowledge of gene expression in bacteria is limited and is based largely on one of the most ingenious generalizations of the century: the operon theory of Monod and Jacob.

Some bacteria do not normally use galactose but they can be induced to do so. Therefore, their DNA must contain the genes that code for the proteins that are needed to use galactose as a

FIGURE 2-3. Schematic representation of human chromosomes.

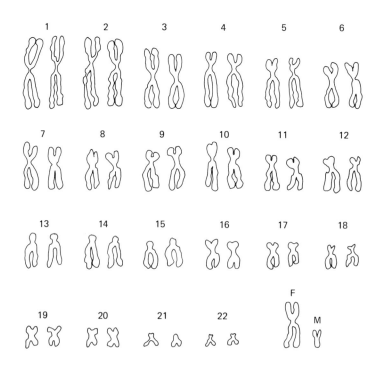

source of food. Three enzymes are needed for galactose metabolism: a permease, a β-galactosidase, and thiogalactoside transacetylase. β-Galactosidase catalyzes the hydrolysis of lactose. The permease transports the lactose into the cell. The role of the thiogalactoside transacetylase is still under investigation. The genes coding for these enzymes are arranged in a rigid sequence, called the operon, in bacterial DNA. In addition to the three structural genes that code for the three enzymes mentioned, there are two other genes: a regulatory gene and an operator gene. A protein, the "repressor," coded for by the operator gene is strongly attached to the repressor gene. When bacteria are grown in the presence of galactose, galactose binds to the "operator" gene and the repressor is detached from it. As a result the operator gene is activated, the structural genes are transcribed into a polycistronic messenger RNA, and the messenger RNA is translated into the three enzymes needed to metabolize galactose. This description of the activation of the galactose operon is greatly oversimplified, and recently a number of other factors have been found to be involved in the activation of the operon.

The structure of the mammalian chromosome is much more complex. Although the molecules regulating gene expression have not been identified, some are suspected. The chromosome is made of a double DNA strand, some RNA, whose role has not been identified, and two major groups of proteins. One group contains

FIGURE 2-4. Schematic representation of human sex chromatin.

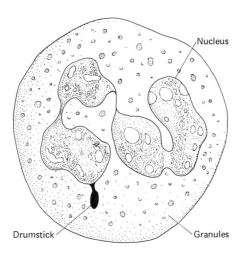

a high proportion of basic amino amids and is called basic proteins or histones, and the other contains more acidic amino acids and is called acidic proteins or nonhistone proteins. Both types of proteins may undergo metabolic changes such as phosphorylation, acetylation, or methylation. The exact relationship of these changes with their potential role in gene expression is not known.

Histones are different from the specific repressors found in bacteria, and there seem to be too few of them to explain the complexity of gene expression in mammalian cells. There is more variability in the nonhistone proteins, and therefore they have been implicated in gene expression. Gene expression is modulated in many aspects of cell life such as enzyme induction, hormone responses, differentiation, and replication. But it may also be involved in some forms of physiological death. For example, the precursor (erythroblast) of the red cell (erythrocyte) is a plump cell with a large nucleus, visible nucleoli, and a rich cytoplasm that makes hemoglobin as the cell matures. During maturation, the cell sacrifices potential diversity of function for efficiency. Thus, the cell progressively represses the expression of its genome and finally (at least in mammalian cells) rejects it altogether. At this stage of its life the red cell becomes a sort of robot with a limited life span and restricted function. Its primary function is exchanging oxygen and CO_2; this process of gas exchange is indispensable for respiration in every cell (Fig. 2-5).

Maturation which culminates in repression of the genome and programmed death is not unique to the red cells. Other cells such

FIGURE 2-5. Conversion of red cell precursor to erythrocyte.

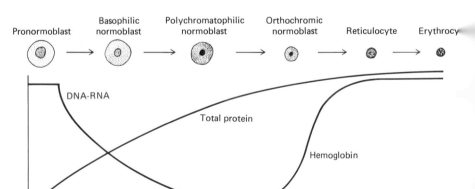

as the epithelial cells of the skin during maturation progressively restrict the expression of their genome to the synthesis of a protein called keratin. The nucleus that was plump and active in the precursor of the skin cell becomes shrunken and condensed in the mature cell. The days of the keratinized cell are numbered and the cell is soon sloughed off.

Whenever the balance between life and death is distorted, some form of disease develops. For example, if red cells die too fast, anemia develops. Psoriasis (although its pathogenesis is not known) may result from a disruption of skin cell maturation.

Gene expression is modulated also in some forms of reaction to injury, for example, liver regeneration. Some cells divide constantly (e.g., the cells of the bone marrow or those of the epithelia of the skin and the gastrointestinal tract). In those cells the genes that code for the proteins involved in DNA synthesis are derepressed. Other cells such as the neurons of the brain never divide, and thus the genes that code for the proteins involved in DNA synthesis are repressed. Other cells, such as liver cells, divide only after special stimuli. When three-quarters of the liver is excised (partial hepatectomy), the remaining quarter grows rapidly and within a few days the normal cell population of the liver is restored. Thus, the cell whose DNA codes for the proteins needed for DNA synthesis, which is repressed before partial hepatectomy, becomes derepressed after the operation. As a result the hepatic cell synthesizes new enzymes, such as: (1) cytidylic and adenylic reductases, which convert ribose nucleotides to the deoxyribonucleotides needed for DNA synthesis; (2) thymidine synthetase, which converts uridylic acid to thymidylic acid, another precursor of DNA; and (3) DNA polymerase, which is indispensable for DNA replication. In addition, the cell synthesizes several other proteins including histones and acidic proteins.

The mechanism that triggers DNA synthesis after partial hepatectomy is unknown. A number of possibilities can be contemplated: elimination of inhibitors of DNA synthesis, the appearance of humoral factors that stimulate DNA synthesis, or a combination of both.

Liver regeneration is an excellent model for the study of regeneration, but other tissues are also capable of partial regeneration, including the kidney after unilateral nephrectomy and the pancreas after destruction with ethionine or partial surgery. We shall see that most hormones also manifest their effect by modulating gene expression. Thus programmed death, regeneration, and hormone action are examples of modulation of gene expression. Inborn errors of metabolism and cancer are manifestations of distortion of gene expression. These few examples emphasize the need for a better understanding of gene expression in mammalian cells.

CELL STRUCTURE

Molecular Organization

The cell is made of many small molecules that are integrated to form larger molecules called macromolecules. Purine and pyrimidine bases combine with ribose or deoxyribose sugars to form ribonucleic or deoxyribonucleic acid, two types of macromolecules involved in gene expression and protein synthesis.

Amino acids are united by peptide bonds to form proteins that play key roles: (1) in cell structure, (2) as catalysts for the conversion of one type of molecule into another (enzymes), and (3) as messengers that convey information from one organ to another (hormones).

Lipids, proteins, and carbohydrates are complexed to form various kinds of cell membranes that prevent entry of undesirable compounds, allow excretion of waste products, are the site of secretion of compounds needed by other cells, and are capable of expanding to move cells or phagocytize foreign elements.

DNA, RNA, histones, and various other types of proteins are complexed to form chromosomes. These are the storage site of genetic information and also are computers programmed for the accurate expression of a portion of the cell's genes (genotype) into a specialized type of cell (phenotype).

A number of molecules—succinic dehydrogenase (a flavin enzyme), cytochrome c, cytochrome c_1, cytochrome b, coenzyme Q, cytochrome a, and cytochrome a_3—are arranged in a rigid sequence to form the electron transport chain, a molecular sequence indispensable for the provision of aerobic sources of cellular energy. These are only a few examples of macromolecular complexes found in cells (Fig. 2-6).

Macromolecular complexes are further organized to form cellular organelles. Little is known of the arrangement of molecules in cellular organelles, and even less is known of the forces that cause macromolecules to assemble in the special shape of organelles. The molecular structure of cellular organelles is too complex to allow study of their assemblage. Nevertheless, preliminary information that will certainly add to our understanding of the buildup of more complex living structures has been obtained on the tobacco mosaic virus (TMV), which infects tobacco leaves. Like all viruses, TMV is made of a core of nucleic acids and a protein coat. The coat is a polymer of a single polypeptide molecule called the capsid. For reasons too complicated to review here, each capsid is shaped like a piece of pie. Three capsids spontaneously combine to form a building unit. Two are placed side by side and one lays on top in the middle of the triangle formed by the two others. These building units are assembled to form a cylinder of two-layered disks, with each disk made of two superimposed layers of capsids and a number of individual capsids. The cylinder resembles a miniature tower of a medieval castle made of pie-shaped bricks (Fig. 2-7).

Although we do not know how cellular organelles are made, we can describe the ultrastructural appearance in some detail. Instead of describing each specialized structure observed with the electron microscope, we will present only an overview of the ultrastructural appearance of the cell (Fig. 2-8). The more specialized structures will be discussed as needed.

Cells are round, cylindrical, polygonal, or oval. Their borders, the plasma membranes, under the electron microscope (after proper treatment) resemble railroad tracks with inner and outer dense lines separated by wider, central, less-dense paths.

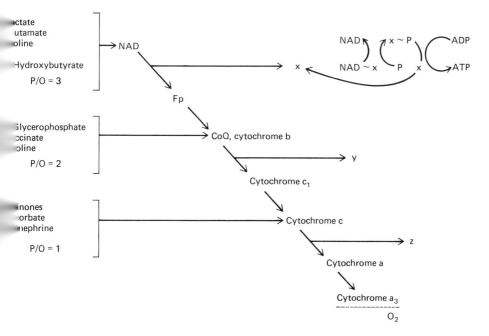

FIGURE 2-6. Electron transport chain and oxidative phosphorylation.

Nucleus

In the center of the cell is the nucleus. Because of its electron density it stands out like the pit of a cut fruit. The nucleus of human cells contains 46 chromosomes, but they can be identified only during mitosis. A distinct round structure in the nucleus is the nucleolus. Both nucleus and nucleolus are important in the transfer of genetically stored information. Such functions will become clearer later.

The nuclear membrane resembles the cell membrane except that it is interrupted to form small holes called pores. Although the function of these pores is not known, it is assumed that they are channels of communication between nucleus and cytoplasm. Not all of the functions of the nucleus in cellular metabolism are known, but it is certain that the nucleus stores in the chromosome most of the genetic information available to any cell and that the nucleus is programmed to provide each cell with what it needs to play its social role in the organ to which it belongs.

Chromosomes can be distinguished morphologically by their shape and staining properties. For example, chromosomes may form Saint Andrew crosses; some crosses are symmetrical so that the upper arms are exactly as long as the lower arms, whereas other crosses are asymmetrical with different sized upper and lower arms.

FIGURE 2-7. Self-assembly of TMV viral coat protein.

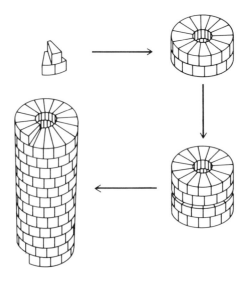

Other chromosomes are shaped like rods and canes. With special techniques using fluorescent quinacrine or Giemsa stain, one can demonstrate a banding pattern unique to each chromosome.

Cytoplasm

The zone between the nucleus and the plasma membrane is the cytoplasm. The cytoplasm is composed of a tridimensional skeleton (the endoplasmic reticulum), a circulating fluid called cytosol, and many smaller organelles (including mitochondria, microbodies, lysosomes, and microtubes), which are often in strategic locations in the cell.

Endoplasmic Reticulum

The endoplasmic reticulum is made of a three-dimensional network of membranes which, with proper methods of preparation, look like the nuclear and cell membranes. The network is believed to connect with the nucleus and the plasma membrane. It is shaped like a distorted beehive. The tridimensional connections are pre-

FIGURE 2-8. Schematic representation of the cell.

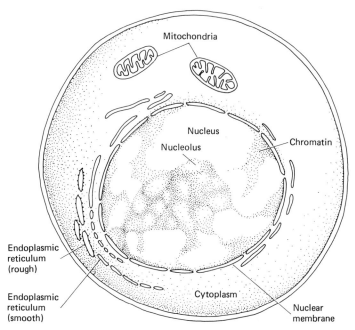

served but irregular. The membranes are studded with small granules, or ribosomes.

Ribosomes are made of ribonucleic acids and proteins, and they play an important role in protein synthesis. Because ribosomes attach to the membranes and give them a bumpy appearance, the portions of the endoplasmic reticulum (ER) membranes covered with ribosomes are called rough, and portions that are free of ribosomes are called smooth. These two forms of ER have different roles in the cell's function. These roles will become clearer later.

Protein Synthesis The ER has numerous metabolic functions, and two principal ones are its roles in protein synthesis and detoxification. Protein is synthesized in the rough ER. The information needed for synthesis of a protein with a specific amino acid sequence is stored in a segment of the DNA molecule referred to as a structural gene. The messenger RNA (mRNA), which is a mirror copy of a DNA structural gene, is transported (sometimes in an extended form) to the cytoplasm by unknown mechanisms; in the cytoplasm it attaches to ribosomes and is translated with the help of enzymes, coenzymes, and RNA into a protein with a specific sequence. The code for DNA transcription to mRNA and mRNA transcription into protein is well known. A specific sequence of three purine, pyrimidine, or a combination of purine and pyrimidine bases dictates what amino acid should be included in the protein sequence. For example, when adenylic acid, adenylic acid, adenylic acid (AAA) is included in the mRNA sequence, a lysine residue is included in the protein sequence.

The translation of the base sequence of mRNA into a protein sequence is not a phenomenon that can occur by chance like the meeting of HCl and NaOH molecules to form H_2O and NaCl, no more than a French sentence can be translated into an English sentence by pulling the words out of a hat one by one. The exact meaning of the words as well as the sequence must be preserved. To ensure that the sequence of the bases in the mRNA is appropriately translated into the protein sequence, the cell has built a sophisticated supporting device, the ribosome. Thus, a ribbon of mRNA binds to several ribosomes to form the scaffold on which proteins are made. A string of five or more ribosomes linked by a ribbon of mRNA forms a polyribosome.

The middleman between the messenger and the protein is transfer RNA (tRNA). It also binds to the ribosomes. Transfer RNA recognizes the codon (e.g., AAA) on the messenger and brings the appropriate amino acids (e.g., lysine) in front of it. Then in a sequence of steps the amino acids are linked together by special enzymes and the new protein is formed (Fig. 2-9).

Chargaff, Crick, Watson, Wilkins, and Kornberg have demonstrated the structure of DNA. In spite of its great significance, DNA is a rather monotonous macromolecule. It is composed of only four different bases (two purines and two pyrimidines): adenine, guanine, thymine, and cytosine. Each base is connected to a phosphorylated deoxypentose sugar forming a long chain called the polynucleotide. The combination of base, sugar, and phosphate is a mononucleotide. Imagine a group of old-fashioned dancers—the men on one side and the women on the other, all holding hands. Men and women may wear one of four colors of clothes, let's say red, blue, green, and yellow. The rows are arranged so that a blue outfit is always opposite a yellow one, and green is always opposite red.

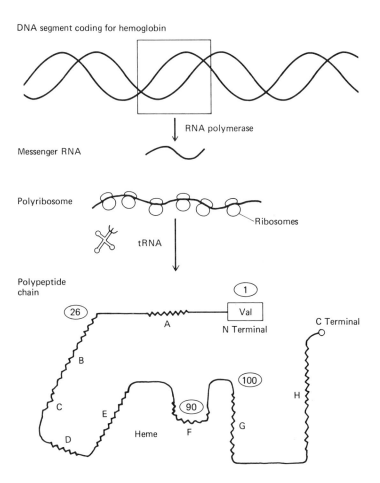

FIGURE 2-9. Schematic representation of protein synthesis.

The pairing of green and red and yellow and blue is never altered throughout the dance. The arrangement of the bases in the DNA molecule is analogous to that of the imaginary dancers. The DNA molecule is composed of two opposing polynucleotide chains in which a thymine residue always pairs with an adenine residue and a guanine residue always pairs with a cytosine residue (Fig. 2-10). The two polynucleotide chains do not face each other like the two rails of a track. Once they have been bound together, they are twisted into a double helix with a highly regular pitch (Fig. 2-11). Inside the nucleus the helix does not extend like a rope stretched from one pole to another, neither is it neatly folded. It is randomly coiled like a long garden hose left to curl on the lawn.

The structures of proteins are more versatile than those of DNA and RNA. Proteins are made of 21 different units called amino acids. The basic skeleton for each amino acid is

but the skeleton is adorned with other atoms of oxygen, carbon, nitrogen, and in a few cases sulfur to form various molecules, each with somewhat different properties. For example, some amino acids attract water (hydrophilic), others repel water (hydrophobic); some carry no electric charge, and others are charged. Those that contain reduced sulfur atoms (-SH) may, after oxidation, bind to other

FIGURE 2-10. Base pairing in DNA.

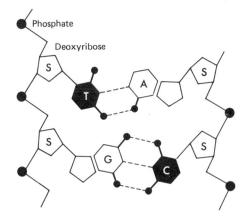

amino acids with reduced SH groups to form covalent -S-S- groups. Amino acids are bound by peptide bonds (-OC-NH-) to form long chains called polypeptides.

The amino acid sequence varies in each polypeptide chain. Again, the polypeptide does not usually form a straight chain, but because of interaction between the amino acids that form its basic structure, the chain is folded. For example, hydrophilic amino acids seek water at the surface, whereas hydrophobic amino acids tend to be tucked inside of the folded polypeptide. Since the combination of amino acid sequences is enormous, the conformation of proteins can vary greatly, as can their functions in the cell economy.

Once molecular biologists learned about the structures of DNA and proteins, they still had to explain how a polynucleotide composed of four bases could be translated in a polypeptide made of 21 amino acids. This problem is analogous to translating a code based on a four-letter alphabet into one composed of 21 letters.

The study of spontaneous alterations (mutations) in the protein

FIGURE 2-11. Watson-Crick model of DNA.

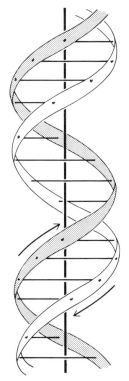

sequence of hemoglobin and of provoked mutations in the proteins associated with the nucleic acid of viruses provided some clues about the amino acid code. It was established that the insertion of one amino acid into a polypeptide chain was dictated by three bases in the polynucleotide chain arranged in a specific sequence (triplet). Since there are four bases in DNA, their combination can yield 64 triplets, yet there are only 21 amino acids. Therefore, it is fair to ask if each triplet is used in the code? We now know that they are, with two exceptions (Table 2-1).

TABLE 2-1. Various codons for each amino acid

Alanine	GCA	GCG	GCC	GCU		
Arginine	AGA	AGG	CGA	CGG	CGU	CGC
Asparagine	AAC	AAU				
Aspartic	GAC	GAU				
Cysteine	UGC	UGU	UGA			
Glutamic	GAA	GAG				
Glutamine	CAA	CAG				
Glycine	GGA	GGG	GGU	GGC		
Histidine	CAU	CAC				
Isoleucine	AUA	AUC	AUU			
Leucine	CUC	CUU	UUA	UUG	CUA	CUG
Lysine	AAA	AAG				
Methionine	AUG					
Proline	CCA	CCG	CCC	CCU		
Serine	AGC	AGU	UCA	UCG	UCC	UCU
Threonine	ACA	ACG	ACC	ACU		
Tryptophan	UGG					
Tyrosine	UAC	UAU				
Valine	GUA	GUG	GUC	GUU		
Ochre	UAG					
Umber	UAA					
Amber	UGA					

Through ingenious and arduous work, Nierenberg and Ochoa discovered the "biological Rosetta stone" and showed that at least 62 of the triplets are used in the code. For example, the sequence CGA in DNA is transcribed into three complementary bases in mRNA (e.g., GCU), the codon. The codon is like a small piece of a puzzle that will fit only with a complementary piece, the anticodon (CGA).

RNA is also a polynucleotide made of four principal bases: adenine, guanine, cytosine, and uracil (instead of thymine). RNA differs from DNA by the presence of a ribose instead of a deoxyribose sugar in each nucleotide. Messenger RNA is essentially a single-stranded complementary copy of the DNA on which it is synthesized by enzymes (RNA polymerases).

The anticodon is made of three complementary bases placed in a strategic position on the tRNA. Transfer RNAs are relatively small polynucleotides folded in a special way. Their conformation permits them to react with a number of other macromolecules. Prominent among these reactions is the coupling of the anticodon with the codon of mRNA at one site and the binding of tRNA with special enzymes that attach a specific amino acid to a specific tRNA on another site.

As mentioned, the messenger ribbon is bound to ribosomes, and the tRNA carrying the amino acids lines up on ribosomes, thus facilitating the elaboration of the polypeptide chain by the formation of a peptide bond between two adjacent amino acids. Once the polypeptide chain is completed, it is shed from the polyribosomal scaffold and folded into its proper conformation. The completed protein can then start to play its specific role in the cell's economy.

Detoxification The endoplasmic reticulum also contains enzymes that detoxify endogenous and exogenous compounds which are or could be toxic to the cells. For example, some ER enzymes conjugate bile or steroid hormones to glucuronic acid and thereby render them soluble so that they can be excreted through the appropriate channels. Other ER enzymes hydroxylate drugs and chemical compounds and often but not always make them inoffensive. The enzymes involved in such hydroxylation are called mixed-function oxidases. Their activity is linked to that of a special electron transport chain which includes cytochromes b_5 and p_{450}.

The Cytosol and Glycolysis

Detoxification and protein synthesis are only two of the many chemical reactions that take place in the cytoplasm. Nucleic acids, steroid hormones, glycogen, fatty acids, and many other compounds are synthesized in the cell cytoplasm. The biochemical pathways involved in these biosyntheses are described in specialized textbooks. These anabolic functions require energy. The two main sources of energy in the cytoplasm are anaerobic (glycolysis) and aerobic (the Krebs' cycle and the electron transport chain) processes. The former do not require oxygen, whereas the latter do.

Glycolysis converts glucose into lactic acid; the Krebs' cycle converts pyruvic acid, some amino acids, and the product of fatty acid degradation into CO_2 and water. In either case adenosine triphosphate (ATP) is formed, and this molecule stores chemical energy in the form of a pyrophosphate bond. The chemical energy can in turn be used to synthesize other molecules to perform work through muscle contraction, or in some case to generate light as in the firefly, or electricity as in the electric eel.

Mitochondria and Aerobic Pathways

The electron transport chain is integrated in a large macromolecular system, the mitochondrion, sometimes called the power plant of the cell because it is the source of cellular energy generated aerobically. The aerobic pathway for generation of ATP is much more effective than the anaerobic: 38 molecules of ATP per molecule of glucose are made available through the first, whereas only 2 are made available through the second.

ATP can be formed through the aerobic pathway only if the sequence of steps involved is rigidly integrated with the electron transport chain. In addition, electron transport must be coupled with the conversion of adenosine diphosphate (ADP) to ATP so that the energy generated by electron release can be transported and stored in the ATP molecule. Tight coupling of electron transport and biosynthesis of ATP is called oxidative phosphorylation.

If a Molotov cocktail is exploded close to a four-wheel vehicle, the vehicle is likely to be moved, but this is hardly an effective way to use chemical energy (gasoline) to displace an automobile. The controlled explosion of the gasoline in the cylinder of a well-tuned engine allows the driver to control the speed and the direction of the displacement of the automobile. Similarly, for chemical energy (glucose) to be converted effectively into a more usable form (ATP), the process must be channeled so that all the steps of the conversion occur in sequence. As was the case with protein synthesis, the ordering of the sequence of events is secured by the special arrangement of the macromolecules. Even minimal interference with macromolecular structure inhibits electron transport and oxidative phosphorylation, and we know a little about the molecular arrangement used for electron transport.

The mitochondrion is also responsible for many other functions, including fatty acid synthesis and oxidation. Little is known of the exact localization of each of these functions inside the mitochondrion. Fatty acid oxidation occurs in the mitochondrion, but it has not been possible to reconstruct the exact pattern of the integration of each enzyme involved in fatty acid oxidation within the mitochondrial structure. Available information suggests that although these enzymes are not freely soluble within the mitochondrion, they are not as tightly bound to the mitochondrial structure as are the enzymes of the electron transport chain.

The general appearance of mitochondria has been described in great detail by electron microscopists. Mitochondria are peanut-shaped structures that are surrounded by a membrane resembling the plasma membrane. The membrane projects infolds inside the lumen forming cristae mitochondriales. Mitochondria are believed to be self-replicating particles. They contain DNA, RNA, ribosomes, and a number of special enzymes that catalyze the biosynthesis of at least some mitochondrial proteins. Mitochondrial DNA does not code for all proteins found in mitochondria; cytochrome c and β-glucuronidase are synthesized in the endoplasmic reticulum. Mitochondrial DNA is believed to code principally for proteins that form the structural framework of the mitochondria.

Other Cytoplasmic Organelles

There exist a number of other intracellular organelles. Some, such as the zymogen granules of the pancreas, have a known function; the role of others, such as the lysosomes and peroxisomes, is not always clear. These organelles will be more appropriately described as we discuss some pathological manifestations of the cell.

Cell Membrane

The cell cytoplasm is separated from the exterior by the cell periphery. Instead of being a wall that separates the interior of the cell from the exterior, the cell periphery is more like a border station and a communication network: it regulates the entrance and exit of molecules. Thus, the membrane exhibits selective permeability, and establishes lines of communication not only with the exterior but also with the interior. Although it has long been known that the membrane is made of lipids (glycolipids and phospholipids) and proteins, only in the last two decades has some insight into the macromolecular arrangement of cell membranes been gained. The membrane is made of a double layer of lipids, which was at first thought to be rather rigid but is now known to be fluid. Within the double layer one finds cholesterol and glycolipids. The fluidity of the lipid layer changes with its composition. For example, the presence of certain fatty acids or increased concentrations of cholesterol increases the rigidity of the lipid layer. Figure 2-12 presents a schematic representation of the principal components of the cell membrane.

Proteins relate to the lipid bilayer in three ways: they may be deposited at the surface of the membrane; they may be partially embedded in the outer lipid bilayer with a portion emerging at the surface; or they may traverse the membrane with a hydrophobic end outside and a hydrophilic end inside. Proteins that seek water at one end and repel it at the other are called amphipatic (Fig. 2-12). The model for an amphipatic protein is glycophorin, which is found in the membrane of the red cell. The fluidity of the cell membrane permits the protein to move within the lipid bilayer, and it has been said that "protein icebergs float in a sea of lipid." The movement of proteins at the surface of the cell is important to the reactions of cells with hormones and antibodies.

A sort of endoskeleton attached to the inner aspect of the membrane is composed of microtubules and microfilaments. We will discuss these structures later; suffice at this point to emphasize their role in membrane movements during cell migration, pinocytosis, phagocytosis, and other processes.

Although most of the proteins in the cell membrane have not been isolated and identified, something is known about their function. One can distinguish various groups of functional proteins: structural (glycophorin) carrier proteins that transfer nutrients (glucose, amino acids, etc.), enzymes that catalyze biochemical reactions, messengers that transfer information, and hormones or antibodies that bind at the surface of the cell to neutralize antigens. Typical among the membrane enzymes are the ones involved in pumping sodium outside and potassium inside of the cell against concentration gradients (Na^+-K^+ ATPase), and adenyl cyclase, which converts adenosine monophosphate (AMP) to cyclic AMP. Cyclic AMP functions as an intracellular messenger for hormones. The integrity of the cellular milieu requires that the intracellular

FIGURE 2-12. Schematic representation of cell membrane.

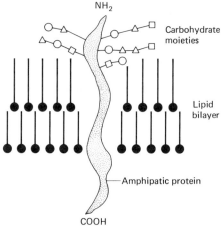

concentration of K^+ be greater than the extracellular concentration, and that the intracellular concentration of Na^+ be smaller than the extracellular. Many other enzymes are found in the cell membrane, but their function is often obscure. The surface of the membrane is covered with macromolecules, probably glycoproteins, which have special affinity for certain compounds, such as hormones, drugs, antibodies, or lectins. These special macromolecules are called receptors, and they will be discussed further when we consider the reaction of cellular antigens with antibodies, the effects of hormones, and the conversion of normal into cancer cells.

INTERCELLULAR INTEGRATION

Microscopic Features

For the human organism to perform adequately, the function of each cell must be integrated with that of the organism. This is achieved by direct communication from one cell to another and by systemic networks of communication.

Except for some biophysical and morphological evidence that cells interact with each other, we know little about what cells tell each other. But if microelectrodes are placed in two liver cells, an electrical current passes from one cell to another. This proves that each cell is not metabolically independent. Moreover, electron microscopists have described a number of anatomical structures that connect the cells, including desmosomes, which seem to hold epithelial cells together, tied junctions, and gap junctions.

Most of the epithelial cells—those sheets of cells that form the skin and cover the surface of the gastrointestinal and respiratory tracts—are hooked together, not by a continuous glue molding the surface like the cement of a brick wall, but by desmosomes, joints that are specifically located at the membrane surface. Desmosomes were long considered to be sites of focal thickening of each membrane, but electron microscopy has disproved this. At the desmosomes two normal bilayer membranes are separated by a normal intermembrane space, which is filled with an electron-dense material probably proteolytic in nature. For unknown reasons, a similar electron-dense condensation is found at the internal aspect of the membrane exactly opposite the membrane junction.

Some cell contacts, the tight junctions, involve the complete fusion of two cellular membranes. Under the electron microscope the surfaces of the two cells are closely apposed at the site of the tight junction, and the classic appearance of the two electron-dense lines of the bilayer membrane is missing, as is the intercellular space. There are only three electron-dense lines, suggesting that the external layers of the bilayer cellular membranes have fused together.

The function of the tight junctions has been deducted from their anatomic location. They are preponderant where a clear separation between two cell types or between a cell layer and the intracellular fluid is indispensable. For example, after intravenous administration many substances never reach the brain, although they may reach all other tissues. The tight junction is believed to be the anatomic counterpart of the physiological blood-brain barrier. Thus, cells that line the blood vessels of the brain are held together by tight junctions; such junctions are not seen in cells that line blood vessels anywhere else in the body.

The third type of junction, the gap junction, forms channels between two cell membranes, penetrating intercellular spaces and connecting the cytoplasm of each cell. Gap junctions are probably critical to communication from cell to cell, but their precise role is not known. The incidence of gap junctions varies with the developmental level of tissues; for example, gap junctions are present in the blastula of some fish but they disappear at later stages of embryogenesis. They are also believed to channel the electrical communications between cells. These conclusions emerge from experiments with the electric organ of the electric eel.

Three major systemic intercellular modes of communication have been developed in the course of evolution: the nervous system, chemical messengers, and blood circulation.

Nervous System

If by mistake one grabs a hot object, the sensory nerves in the skin of the hand convey a message to brain neurons, which transfer the information to other neurons to which they are connected, and these neurons finally convey their message to motor endings in muscles. As a result, some muscles contract to open the grasping hand and the hot object is dropped (Fig. 2-13).

The central nervous system is composed of the spinal cord and the brain. Its function is to perceive sensory impulses, emit motor impulses, and store information in the processes of learning and remembering. Thus, the senses of smell, taste, touch, hearing, and sight function through sensory nerves which are connected to brain cells. The brain cells emit impulses that are transferred to the muscle: of the tongue and larynx to speak, of the thorax and diaphragm to breathe, of the esophagus to swallow, etc.

The cells controlling the most delicate functions of the nervous system (coordination of movement, control of speech, memory) are located in the superficial parts of the brain called the cortex. The cortex is made of many different types of cells, but the most important for nerve function is the neuron (Fig. 2-14). The neuron is unusual in that its body emits fibrillar cytoplasmic extensions called dendrites and axons. These fibrillar structures are wrapped in a sheet of myelin. The myelin is an extension of the membrane of a cell other than the neuron (the so-called Schwann cell in the peripheral and the oligodendrocyte in the central nervous system). Like the cell membrane, myelin is made of proteins and lipids. We know practically nothing about myelin synthesis and function. Yet when myelin is not made, neurons do not function properly.

Chemical Messengers

Glucose is a major source of cellular energy, yet it cannot penetrate peripheral muscle membranes. After glucose administration, the β cells of the pancreas are stimulated to secrete a chemical messenger, the polypeptide hormone insulin, which modifies the permeability of the muscle membrane and facilitates the penetration of

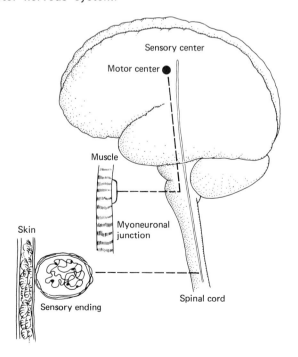

FIGURE 2-13. Schematic representation of sensory and motor nervous system.

glucose. (As shown later, this is a greatly oversimplified view of hormone action.)

Circulatory System

In the course of evolution, organisms made of a multiplicity of cells grew larger. As their number and variety increased, it became impossible for each cell to absorb oxygen or needed nutrients directly from the environment without a circulatory system. In humans, as in most mammals, the circulatory system is composed of channels lined by cells and cellular products. The simplest of these channels, capillaries, are lined by a single layer of endothelial cells resting on a basal membrane made of a special type of connective tissue fiber. The more complex vessels are the arteries, veins, and lymphatics, which in addition to the endothelial wall have various layers of connective tissue and elastic and muscle fibers. The channels contain circulating fluids and cells.

FIGURE 2-14. Schematic representation of neuron.

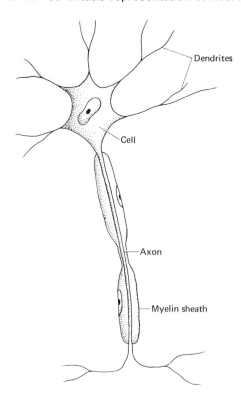

Many proteins as well as smaller molecules are in solution in the circulatory fluids. Some important proteins of the circulating fluid include antibodies and a variety of proteins capable of binding smaller molecules, which can thereby be transported from one organ to another. For example, transferrins transport iron, ceruloplasmin binds to and transports copper. Specific proteins with binding affinities for hormones (e.g., thyroid hormones) are also present in the circulating fluid. Of course, a large number of smaller molecules circulate as well. Among them are free fatty acids, amino acids, some enzymes, and injected or ingested drugs.

The circulatory system carries also four major groups of cells: red cells, whose principal function is to transport oxygen and eliminate CO_2; polymorphonuclears, which play a key role in inflammation; and lymphocytes and monocytes, which are involved in immunological defense. In addition, the circulating blood contains platelets, which are derived from large bone marrow cells called megakaryocytes. Platelets are not exactly cells: they resemble segments of cytoplasm that contain small granules. Platelets and a number of other circulating proteins are crucial to blood coagulation.

The circulatory system is composed of a central pump, the heart, a number of different types of blood vessels—arteries, veins, and capillaries—and a ventilation system, the lungs. The purpose of the circulatory system is to provide every cell of the body with oxygen and to eliminate the CO_2 formed during metabolic conversions. The carrier for the two gases is a major component of the red cell, hemoglobin. Metabolic CO_2 binds to hemoglobin and the red cells carry it from the rest of the body to the heart by veins.

The heart is made of four chambers: a right and left atrium and a right and left ventricle. The veins of the systemic circulation converge to form the superior and inferior vena cava (SVC and IVC). The deoxygenated blood enters the right atrium (RA) and is transferred to the right ventricle (RV). From the right ventricle the deoxygenated blood is pumped into pulmonary arteries and transferred to the lungs (L) where it is reoxygenated. The oxygenated blood is returned to the left atrium (LA), transferred to the left ventricle (LV), which pumps the blood into the aorta, and from the aorta branches emerge that distribute the blood to the rest of the body (Fig. 2-15).

CONCLUSION

In conclusion, in living organisms the atoms are organized in small molecules (purine, pyrimidine bases, amino acids, fatty acids, etc.).

Small molecules are aggregated to form macromolecules, glycogen, polynucleotides, polypeptides, and lipids. Macromolecules may interact and circulate in the blood, but often they form structural units such as chromosomes, nuclear membranes, ribosomes, endoplasmic reticulum, mitochondria, lysosomes, peroxisomes, and cell membranes. All cell injuries ultimately interfere with the macromolecular arrangement of the cell.

FIGURE 2-15. Schematic representation of blood circulation. Ao, aorta.

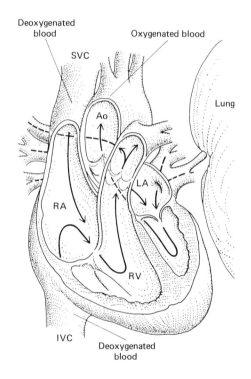

DEFENSE MECHANISMS

INTRODUCTION

Although one can only speculate about the beginnings of life, it seems certain that the environment—the oceans, the earth's crust, the atmosphere, and sunlight—in which the first unit of life emerged turned against the newly created molecules (for instance, nucleic acids). Therefore, persistence of life was a struggle even for the simplest form of life. When the human species appeared, the earth was already inhabited by millions of plants, bacteria, and animals.

To live, man must breathe and eat. Even the air of primitive man was not pure; pollen and smoke have always been present. Moreover, even organic foods were probably contaminated by animal excretions; plant, fungi, and bacterial toxins; viruses; and parasites. Therefore, even the simplest organism needed to develop mechanisms for disposing of particulate matter, nefarious molecules, and biological intruders. In humans particulate matter is eliminated by cilia in the upper respiratory tract, by phagocytosis (a process linked to inflammation), and nefarious molecules are neutralized either through metabolic conversion or by the elaboration of specific antibodies. Biological intruders trigger inflammation and humoral and cellular immunity.

Some injurious agents cannot be stopped before they reach and alter important molecules. In such cases the cell may be able to repair the damaged molecule; a case at point is the change in DNA caused by ultraviolet light. UV light causes the formation of covalent bonds between two adjacent thymine residues, thereby distorting the DNA molecule. The cell has devised an enzyme mechanism whereby it can excise the dimers and patch up the DNA strands in what is sometimes called the cut-and-patch repair mechanism. Other forms of DNA repair exist, one occurs during DNA replication (postreplication repair). Macromolecular structures other than DNA might also be repaired. For example, collagen fibers damaged by drugs, such as hydralazine, are removed by digestion, and this is followed by reconstruction of the fibers. When liver cells are separated and placed in culture medium, they are leaky. However, after 5 days they recover, which indicates that their membrane must have been restored.

As life evolved it could be sustained only by life, and as a result living organisms competed fiercely for survival. To protect themselves from an environment that was not totally hostile but never entirely friendly, living organisms built powerful barriers between their beings and their surroundings to maintain their equilibrium or homeostasis.

FIRST LINE OF DEFENSE

In the human organism the first line of defense is secured by the epithelia of the skin and of the gastrointestinal, respiratory, and genitourinary tracts. When these barriers are interrupted by spontaneous weakening—e.g., in the presence of a stomach ulcer (Fig. 3-1) or damage caused by environmental agents—the breach may threaten the entire organism by loss of blood or by permitting invasion by other living organisms such as parasites (e.g., hookworm), bacteria (as in tetanus), viruses (as in the common cold or poliomyelitis), or fungi (as in athlete's foot). To survive even minor local disruptions of its protective barrier, the organism has developed several major mechanisms of defense: blood coagulation, inflammation, and immunity. Injuries of various kinds may lead to the loss of substance not only of the protective barrier, but also of the deep-seated tissues such as the lungs, liver, or pancreas. To restore the lost substance, the organism resorts rarely to regeneration (or integral restoration of the lost cells) and more frequently to wound healing.

BLOOD COAGULATION, INFLAMMATION, AND IMMUNITY

The events that result in blood coagulation, inflammation, and immunity have four common properties: (1) they occur in several steps; (2) they involve the interaction of molecules in the circulating plasma, in intercellular tissue, and in cells; (3) one or more steps of the sequence are carefully regulated to prevent their spontaneous unleashing or continuous manifestation; and (4) some of these defense mechanisms occur very rapidly because they are autocatalytic, i.e., the rate of the full deployment of the event is at some stage fantastically amplified. Nonautocatalytic and autocatalytic schemes are presented in Figure 3-2.

In the nonautocatalytic scheme, all the steps of the reactions occur in rigid sequence and each molecular event leads to the next.

In the autocatalytic scheme, the sequence is also orderly, but at step 2, for example, the event becomes autocatalytic and is thereby multiplied.

Thus, let's assume that: (1) an inactive factor before step 2 needs a factor x for activation to 2; (2) x is generated from z; (3) 2 converts z to x; and (4) 2 is a precursor of 3. Because in this scheme substance 2 catalyzes its own production, the conversion of $2 \longrightarrow 3$ is extensively multiplied over that of $1 \longrightarrow 2$. The process can be further accelerated if one or more steps of the sequence are also autocatalytic. Such autocatalytic events occur in the chain of events that take place in blood coagulation, inflamma-

FIGURE 3-1. Schematic representation of a gastric ulcer.

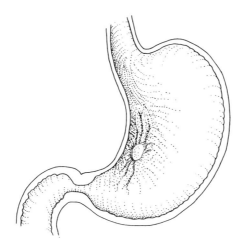

tion, and immunity, and they are responsible for the rapidity of the reactions.

The cascade of defense reactions involves the overlapping of activation of inactive molecules and inactivation of active molecules. This complex overlap of antagonistic chemical reactions, coupled with the rapidity of the reactions, explains why some of the events in blood coagulation, inflammation, and immunity are so difficult to understand.

The active compounds are usually proteins, but they exist in an inactive form. The purpose of the activation is to bring the active center of the protein molecule in contact with its target. This can be achieved in at least two major ways: a change in protein conformation from an inactive to an active one, or the release of an entrapped active protein.

Many of the studies on protein activation have been done on enzymes, but what has been learned from enzymes applies also to those proteins involved in defense mechanisms. In fact, as we shall see, most of the proteins involved in defense are enzymes.

The protein that constitutes the enzyme may be inactive and may require for activation a prosthetic group. Prosthetic groups are usually small molecules such as phosphate, iron, or vitamins that bind to the protein (usually after appropriate metabolic conversions).

Often enzyme activity is modulated by substances that inhibit or activate it. Such modulation of enzyme activity may occur either at the active site or at some distance from the active site (the allosteric site).

Insulin is made of A and B polypeptide chains which are held

FIGURE 3-2. Autocatalytic and nonautocatalytic defenses.

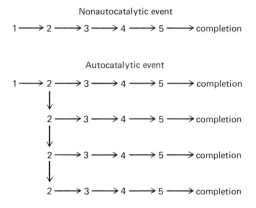

together by S-S bonds. The precursor of insulin is an extended polypeptide which contains the A and B sequences connected by a C sequence, or connecting polypeptide; activation of the hormone requires that the connecting peptide be oxidized.

The protein conformation of many proteases contains a chain of amino acids that form a histidine loop. Phosphorylation of one serine residue included in the loop results in a conformational change sufficient to convert an inactive into an active protease.

Metals are often indispensable to enzymic action. They usually are bound to or close to the active sites either by simple ionic bonds or by the more complex process of chelation. When a metal is chelated, it is entrapped between two arms extending from the body of the molecule like the prey is caught between the claws of a crab.

Other molecules are activated in a completely different fashion. The protein is elaborated in the form of an extended precursor; i.e., the polypeptide chain is much longer than it should be to display activity. Activation takes place by splitting portions of the molecule. For example, during blood coagulation the fibrinogen molecule is split at four different points, generating two fibrinopeptides (since the molecule is a dimer). The cutting of the two fibrinopeptides from the fibrinogen molecule yields a molecule of fibrin that is apt to polymerize with other molecules of fibrin and therefore plays a key role in blood coagulation.

Trypsinogen and chymotrypsinogen are activated only after excision of a portion of their polypeptide chains. Insulin and several other hormones are not made directly by the cell, but precursors with an extended polypeptide chain are formed first. A special enzyme splits the precursor to yield the active hormone.

Sometimes the inactive molecule is strapped by covalent (e.g., S-S) or ionic bonds to membranes and becomes active only after these bonds are ruptured by special stimuli.

Finally, some active molecules are kept inside of a closed envelope and thereby prevented from acting on their immediate environment. Histamine, a powerful vasodilator, is contained in such granules and only after rupture of the granule is the chemical mediator released and able to exert its effect. The various mechanisms of protein activation are presented in Figure 3-3.

Inactivation may also occur in various ways: metabolic conversion (drugs, carcinogens, and hormones), binding of inhibitors to active or allosteric sites (e.g., antimetabolites and nucleotides in orotic aciduria), and formation of inactive molecular complexes that may in turn be degraded. Only the last mechanism of inhibition needs further consideration. A typical example is the interaction between antigens and antibodies. The antigen binds to specific sites on the antibody forming a complex that may be either excreted or precipitated and phagocytized.

Blood Coagulation

Now that we are somewhat more familiar with the types of molecular events involved in the various defense mechanisms, these processes can be described briefly.

FIGURE 3-3. Various mechanisms of protein activation.

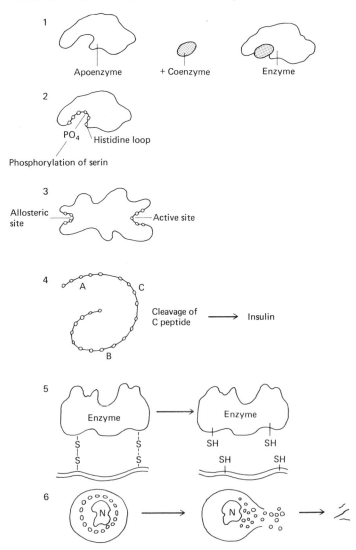

The purpose of blood coagulation is to secure hemostasis, or to prevent the loss of blood. When the continuity of the vascular system is interrupted, at least three conditions are required to make blood coagulation successful: (1) coagulation must occur soon after the injury, (2) it must not extend beyond the lesion so as not to compromise circulation in vital blood vessels, and (3) it must not be permanent. Thus, under optimal conditions coagulation is self-limited and partly reversible. The steps of blood coagulation are schematically represented in Figure 3-4. Blood coagulation occurs in a triple sequence: formation of thromboplastin, conversion of fibrinogen to fibrin, and platelet aggregation. It culminates in the formation of thrombi.

Thrombi are intravascular bodies formed from soluble and cellular blood components during blood coagulation. When the blood vessel is injured because of disease, such as atherosclerosis or trauma, platelets first assemble at the site of injury in the process of platelet agglutination. In this first stage platelets are held together by weak bonds, but in a second stage they are tightly bound, forming a compact mass that adheres to the vascular wall. In the next stage the normal structure of the platelets becomes blurry; the granules disappear by bursting and releasing a number of chemical mediators, including factors that lead to the appearance of thromboplastin. Thromboplastin converts prothrombin into thrombin, a proteolytic enzyme that converts fibrinogen into fibrin. The fibrin molecules polymerize and form a network that entraps white cells, red cells, and more platelets. If thrombus formation or thrombosis goes unchecked, the thrombus may occlude the vessel. If the vessel

FIGURE 3-4. Steps in blood coagulation.

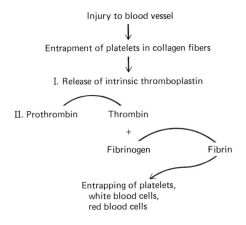

is an artery such as the coronary artery or one of the cerebral arteries, blood flow to the irrigated area is interrupted.

Inflammation

To illustrate the events in acute inflammation, let's consider two familiar lesions: the boil and pneumonia. Almost everyone has a boil at one time or another; pneumonia was and still occasionally is a dreadful disease. The agents responsible for boils are bacteria, almost always staphylococci. If the bacteria find a crevice at the surface of the skin, they penetrate it. The tissue underneath the skin epithelium is rich in foods that help the staphylococci proliferate. As the colony of bacteria grows, it makes some strategic mistakes: the bacteria secrete substances that attract polymorphonuclears from the circulating blood, and they destroy some of the tissue cells that surround them. This triggers the battle between host and parasite. The host responds by increasing the blood circulation and dispatching armies of polymorphonuclears to the infected area to control bacterial growth.

Polymorphonuclears are small white cells that are made in the marrow and circulate in the blood (see Fig. 3.5). The three types of polymorphonuclears—basophils, neutrophils, and eosinophils—differ in their overall appearance and staining properties and can therefore be recognized readily by the experienced pathologist or hematologist.

We know little of the role of eosinophils in inflammation. Basophils contain granules that are released in inflammation. The granules in turn contain chemical mediators, among them histamine. When the granules burst, histamine is released and causes dilatation of the blood vessels, increasing local blood circulation. The spurt of blood generates heat and causes the redness characteristic of the inflamed area. For reasons that we cannot discuss here,

FIGURE 3-5. Schematic representation of blood cells. P, polymorphonuclear; L, lymphocyte; M, monocyte; E, erythrocyte.

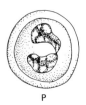

P   L   M   E

the capillaries in the inflamed area become leaky and plasma exudes from the capillary lumen into the interstitial tissue. This accumulation of fluid causes swelling or edema. The distended tissues compress the nerves and cause pain. At this stage the classic hallmarks of inflammation are all apparent: redness, warmth, swelling, and pain. These signs were recognized by Celsus, who stated, *"Signa vera inflammationis sunt rubor calor et dolor."*

Vasodilation also sets the stage for polymorphonuclear invasion. Polymorphonuclears traverse the vascular wall by diapedesis. In most cases it is believed that they pass through gaps between endothelial cells, but they have been seen to pass right through the cytoplasm of the endothelial cell.

Polymorphonuclears are reminiscent of the Japanese kamikaze pilots. As they mature they fill their cytoplasm with granules loaded with hydrolytic enzymes, substances capable of killing bacteria and other invaders. When they approach the bacteria that have attracted them by elaborating chemotactic substances, the polymorphonuclear membrane sends out infolds that entrap the bacteria and ingest them in the process of phagocytosis, from the Greek word meaning "cell eating." Once a group of bacteria are inside the polymorphonuclear they are killed (at least in the case of staphylococci), but the polymorphonuclear dies in the process. The mixture of dead polymorphonuclears, bacteria, and tissues forms a thick, well-bound, yellow "pus" that most of us mistakenly press out (it contains live bacteria). Once the pus is eliminated the boil usually heals.

In pneumonia the nefarious bacteria are often pneumococci. They are inhaled, enter the bronchi, and travel to the alveoli where they secrete their deadly poison. The killed cells, blood, and plasma that leak through the capillaries are marvelous grazing ground for pneumococci. The bacteria proliferate; through chemotaxis polymorphonuclears accumulate; and the capillaries dilate and become leaky. The leakage of plasma in the alveoli and the massive migration of neutrophils ultimately fill the entire alveolar space impairing gas exchange: the uptake of oxygen and the elimination of CO_2. Clearly, at this stage the struggle between host and parasite is critical. If the parasite takes over more territory in the lungs, in spite of the inflammatory reaction, large portions of both lungs are nonfunctional, the host is unable to breathe and dies in respiratory failure. However, if the inflammatory reaction can restrict the bacterial invasion, for example, to one lobe of one lung, all bacteria may be killed, and in time that portion of the lung will be cleared of the infiltrating bacteria (Fig. 3-6).

In chronic inflammation lymphocytes, plasmocytes, and macrophages invade the inflammatory area. Chronic inflammation is not the all-out onslaught between host and parasite that acute inflammation is. By parasite we mean a virus, bacterium, fungus, or complex parasite such as the agent of malaria or even the hookworm.

Chronic inflammation reflects an attempt on the part of the parasite and host to live together (symbiosis). However, for some reason the parasite seems to erode the defenses of the host slowly, and often chronic inflammation causes irreversible damage to the organ affected. The damage is not always due to the parasite's mechanical presence, as in filariasis, or to its elaboration of toxins, but it often results from disproportionate inflammatory or immunological reactions. Because most of the cells found in chronic inflammatory areas are involved in immunological reactions, many of the manifestations of chronic inflammation will be better understood as we learn more about immunity.

At the present we will describe the histological appearance of an area of chronic inflammation.

There are two major forms of chronic inflammation, classic and granulomatous. The common features in both types are discrete parasitic infection, cellular necrosis coupled with an attempt to repair the dead tissues, and infiltration by cells with immunological potential.

In classic chronic inflammation, the infectious agent, whether it be a virus, bacterium, or complex parasite, is not always easy to detect. Yet cellular damage occurs, and as a result of tissue loss parenchymal cells regenerate and fibroblasts proliferate and elabo-

FIGURE 3-6. Schematic representation of lobar pneumonia of right lobe, frontal view.

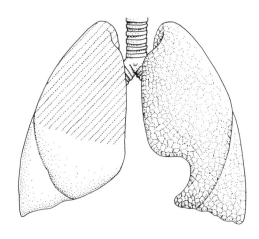

rate new connective tissue fibers. Depending on whether it is sparse or dense, the accumulation of connective tissue is referred to as fibrosis or sclerosis. The invaders are usually antigenic and thereby call in cells capable of mounting an immunological reaction.

The granulomatous reaction is much more complex, and little is known about its pathogenesis. Usually three major types of cells are involved: lymphocytes, epitheloid cells, and giant cells. In addition, a great deal of material often accumulates as a result of cellular and bacterial death. The necrotic material and the cellular components of the granulomatous infection are distributed in tissues in various arrangements depending on the type of infectious agent.

A typical form of granulomatous inflammation is tuberculosis. The bacteria responsible for this infection are much more insidious than staphylococcis. They do not invite a massive attack by polymorphonuclears. They usually infect a young host, settling in the lungs and spreading to the adjacent lymph nodes, which causes minor damage in both organs. The infected child exhibits symptoms no worse than those of a cold, and after a few years only a star-shaped scar and a calcified lymph node appear on the X-ray. Later in life when the host's defenses weaken, the bacteria which are still alive seize the opportunity. They invade the lung tissues, proliferate, and elicit the cellular reaction leading to formation of the tubercle: the unit granulomatous lesion. The tubercle is com-

FIGURE 3-7. Schematic representation of the histology tuberculosis.

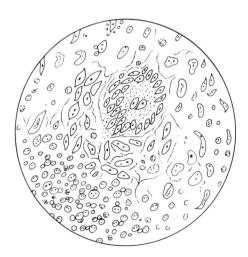

posed of a central core of insoluble necrotic material called caseum (the Latin word for cheese). The caseum is surrounded by large multinucleated cells and a crown of mesenchymal cells that resemble epithelial cells and are therefore called epitheloid cells. At the periphery of the tubercle several layers of lymphocytes are found (Fig. 3-7).

Immunity

One of the most ingenious mechanisms of defense against intruders is the elaboration of antibodies. Antibodies are like policemen tailored for each criminal. Although they are all γ-globulins, they exist in various molecular forms: IgE, IgA, IgG, IgD, and IgM. Each form can perform different tasks (Fig. 3-8).

With respect to their distribution in the body, there are two major kinds of antibodies: those that circulate freely in the blood and those that remain bound to cells. Because of this dichotomy in the immunological mechanism, one speaks of humoral and cellular immunity. It is not certain whether or not these two forms of immunity developed at the same time. However, all precursors of antibody-forming cells are lymphocytes elaborated in the bone marrow. Two different types of mature cells, B and T cells (Fig. 3-9), elaborate humoral and cellular antibodies, respectively. In the chicken the B cell is formed in the bone marrow and is transferred to the bursa of Fabricius where it is programmed to elaborate humoral antibodies (Fig. 3-10). In mammals B cells originate in the bone marrow, but it is not known where they are programmed.

T cells also originate in the bone marrow; they then migrate to the thymus, where they are programmed to elaborate antibodies that remain attached to the cell. Humoral antibodies detect circulating antigens and complex with them, thereby neutralizing them, as is the case for the diphtheria toxin. The antibody may also precipitate the antigen or agglutinate the antigen carrier as in red cells.

The antigen-antibody complex often triggers an inflammatory reaction by activating complement. Serum contains a population of proteins that are activated to interact with each other and yield substances that cause chemotaxis, cellular death, and opsonization. This population of protein is called complement. There are at least 11 different components of complement: C1, C2, C3, C4, C5, C6, C7, C8, C9; C1 is a complex of $C1_q$, $C1_r$, and $C1_s$. The mechanism of activation of complement is too complex to be described here, but suffice it to point out that it is an autocatalytic event reminiscent of blood coagulation.

As we have mentioned, T cells are programmed in the thymus and the antibody remains attached to the cell. This mounted policeman is uniquely suited to patrol all parts of the body. As soon as it detects intruders—which may be bacteria like the agent of tuberculosis, cells of grafted tissues, or even cancer cells—it elicits an inflammatory reaction by elaborating a number of special substances and binding to complement, and it triggers its own proliferation increasing the number of combating cells without the need for new recruits.

In mounting their defense against intruders, the two types of immunological cells, B and T cells, do not work at cross purposes; in fact, they often interact. The molecular mechanisms of these interactions are poorly understood. In addition, T and B cells often call macrophages to their aid.

Thus, the intrusion of foreign antigens—which may be free proteins (tetanus toxin), drugs (penicillin), proteins, or carbohydrate fractions associated with viruses, bacteria, parasites, or foreign cells—elicits a B- or T-cell response, or both. The immunologically competent cells proliferate and may trigger an inflammatory reaction by activating complement and calling in macrophages for assistance. The exact role of macrophages in the immune response remains to be clarified. A schematic representation of the reaction between antigen and antibody is presented in Figure 3-11.

FIGURE 3-8. Molecular structure of IgG.

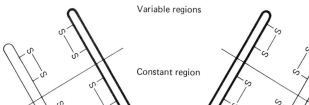

RESTORATION OF LOST TISSUES

In his fight for survival, and also because of his ambition to dominate the world, man got hurt. Arms and legs were broken in falls; animal bites caused wounds almost everywhere; fires burned the skin and sometimes deep tissue; knives and axes cut off fingers, hands, entire limbs, and sometimes the head; arrows and bullets pierced human bodies to damage the internal organs.

Thus, by accident or by design, humans frequently lose substantial amounts of their body tissue. Survival requires that either the damaged tissues be integrally restored or the wound be patched. In prehistoric times repair or wound healing had to take place fast enough for the victim to continue to obtain food and shelter.

Mammals are much less effective than amphibians in repairing lost tissues. When a tadpole's leg is cut it regenerates, and a new limb develops in 3 weeks. Excision of the lens in a salamander is followed by complete restoration of that organ.

Although transplantation of nerves to the stub of the amputated limb of a rat induces some expansion of the remaining stub in what looks like an attempt to regenerate the limb, complete restoration of limbs has never been achieved in mammals. It has been argued that poikilotherms (cold-blooded animals) can regenerate limbs because they can survive longer without food. In contrast, homeotherms (warm-blooded animals) need to restore the lost organs rapidly if they are to find food for survival. It is difficult to

FIGURE 3-9. Origin of T and B cells.

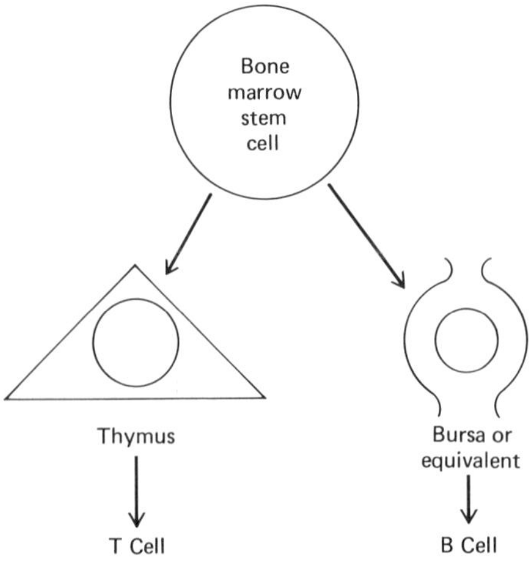

argue for or against such a hypothesis because evidence is unavailable.

Wound Healing

By hypertrophy, hyperplasia, or a combination of both, mammals can adapt some of their organs to increased needs or restore lost substances. They can also repair interrupted tissues, such as skin and mucosae, by wound healing. Wound healing results from the interaction of complex cellular and humoral factors.

To explain wound healing, it is easiest to describe a simple case, for example, the wound caused by the sterile scalpel of a surgeon during an appendectomy. The cut penetrates the epithelial layers of the skin, the subcutaneous connective tissue, the muscles of the abdomen, and the visceral peritoneal wall. All cells that come in direct contact with the scalpel are killed. The incision also interrupts a number of blood vessels. The surgeon controls the bleeding by placing a hemostat at strategic spots. Once the operation is over, the open wound is closed. If too many cells have been killed by the pressure of the hemostat, it is sometimes wise to eliminate with scissors the tissues that have been severely traumatized before suturing the wound. The wound is then sutured by layers: peritoneum, muscle, dermis, and epithelium. The lips of the wound are brought together tightly so that no gap is left, but not so tightly that the lips of the wound overlap.

The damage caused by the scalpel is followed by vascular and cellular reactions. In the beginning vascular reactions predominate. Vasodilation of the capillaries takes place in the live tissues that surround the incision. Vasodilation is associated with leakage of plasma in the interstitial fluid. As a result the skin around the wound is red, warm, and the lips of the wound are slightly swollen.

FIGURE 3-10. Formation of antibodies from B lymphocytes.

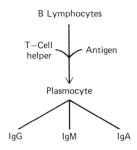

Even after suture, the gap is not completely closed, and some degree of hemorrhage occurs between the lips of the wound. If blood coagulation functions normally, the extruded blood clots.

Cellular death triggers an inflammatory reaction with leukocytic margination and diapedesis. Polymorphonuclears accumulate at the edges of the wound, especially around blood vessels, forming cuffs. Columns of polymorphonuclears move toward the necrotic tissues and phagocytize them. Once hemostasis is established and the dead

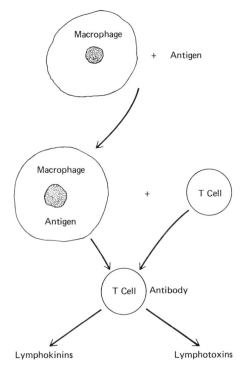

FIGURE 3-11. Schematic representation of reaction between antigen and antibody. The antigen, usually attached to a cell or a bacterium, is processed in some unknown way by a macrophage. The macrophage cooperates with an "uncommitted" T cell to convert it into a "committed" T cell, which then carries antibody molecules to the antigen. The committed T cell secretes lymphokinins that attract uncommitted T cells and sensitize them to the antigen. Other T cells called killer cells secrete lymphotoxins. They have receptors for the antigen and complement and can kill the agent carrying the antigen, sometimes without the help of B-cells, but in other cases with the assistance of B cells producing IgG.

tissues are scavenged, restoration takes place. It starts with the formation of granulation tissue (not to be confused with granuloma formation). The small capillaries at the edge of the wound proliferate actively to invade the blood clot that holds the edges of the wound together. They are accompanied by young fibroblasts which proliferate actively as the clot is progressively digested by polymorphonuclears. This combination of young capillaries, young fibroblasts, and polymorphonuclears constitutes the granulation tissue. Occasionally a few lymphocytes and a few plasmocytes are also seen in granulation tissue.

Epithelial cells of the skin migrate from the edge of the wound toward the center. This migration is not caused by extension of the epithelial layer by cellular proliferation but by actual displacement of the individual cells toward the center of the wounded area. Later the cells that migrated and the cells at the edge of the wound undergo mitosis and restore the entire epithelial layer through cellular proliferation. A similar process repairs the peritoneum.

The muscle fibers are not restored, and the discontinuity in the musculature caused by the scalpel is filled with connective tissue composed of fibroblasts and collagen fibers.

Obviously, factors that interfere with blood coagulation, inflammation, cellular proliferation, and elaboration of connective tissue modulate the process of wound healing. Wound healing is retarded if the trauma has been excessive, the patient is malnourished, or the patient has been under prolonged shock during the operation. When a hematoma (severe bleeding) forms at the site of the wound because of vitamin K deficiency, the lips of the wound are kept apart and epithelial restoration is delayed. Proliferation of connective tissue is impaired by the absence of vitamin C. The administration of X-irradiation soon after the incision is made destroys proliferating cells and prevents adequate wound healing.

One of the most intriguing questions raised by wound healing is what triggers and stops proliferation of epithelial cells and fibroblasts? Two mechanisms could trigger cellular proliferation: elimination of inhibitors present in the intact cell and elaboration of stimulatory substances after injury. Available evidence suggests that the normal epithelium contains inhibitory substances, called chalones, which prevent proliferation. Chalones have been partially purified; they are believed to be polypeptides. However, a great deal of work is still needed to establish their role in wound healing.

Fractures

Most skiers sooner or later fall and break a leg or an arm. Seldom does the pain and the inconvenience caused by such an accident prevent them from returning to the snow-covered slopes, because the interrupted bone is readily repaired by a process akin to wound healing. When the bone is broken one speaks of fracture. Let's assume that an unfortunate skier breaks his tibia. The fracture is clean-cut, the upper portion is separated from the lower portion of the bone, and fortunately the bone is not fragmented and the broken extremity did not slide in such a fashion that it destroyed the surrounding muscle or perforated the skin. Except for the pain and his inability to walk, the skier would not know that he had a fracture if it were not clearly revealed by an X-ray (Fig. 3-12).

The restoration of a fractured bone is somewhat more complex than wound healing. The original bone is composed of a thin layer of connective tissue, the periosteum, which is closely attached to a thick layer of bone called the cortex. In the center of the cortex is a cavity filled with bone lamellae and either fat or active bone marrow, depending on the age of the individual. Immediately after trauma, hemorrhage takes place at the site of the fracture. The blood coagulates and soon the hematoma is invaded by granulation tissue. The proliferation of blood vessels and the presence of

FIGURE 3-12. Schematic representation of fractures.
A. Neck of femur.
B. Humerus.

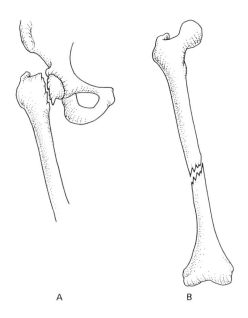

polymorphonuclears in some way stimulate the breakdown (osteolysis) of the free edges of the fractured bone.

At a later stage the granulation tissue is invaded by young fibroblasts that elaborate collagen. The calcium released from bone fragments and the edges of the fractured bone precipitates on the connective tissue and forms the primitive ossifying callus. Such a callus is friable and fragile, and if the bone is to be restored to its original strength, the callus must be replaced. This is achieved by a rigid sequence of morphogenetic steps involving resorption of the primitive callus, its replacement by cartilaginous cells to form the cartilaginous callus, calcification of the cartilaginous cells, and remodeling of the new bone according to pressure forces to yield the pattern of calcification of normal bone.

Thus, in fracture repair, cell proliferation and a strict morphogenetic rearrangement restore the bone to its original structure and strength. The conversion of a primitive cell into a cartilaginous or a bone cell (osteoblast and osteoclasts) and the mineralization of the connective and the cartilaginous tissues have been described in detail morphologically. Yet little is known of the molecules involved in cellular maturation or in triggering mineralization. Interference with the connective tissue proliferation (e.g., vitamin C deficiency) or calcification (hyperparathyroidism or vitamin D deficiency) hinders formation of a proper osseous callus and the restoration of bone continuity.

Hypertrophy and Hyperplasia

Losses of parenchymal tissue can often be repaired by hypertrophy and hyperplasia. Hypertrophy results from the enlargement of individual cells without cell division; in hyperplasia, the proliferation of individual cells, the new cells replace the lost tissue. A number of parenchymal tissues—including some exocrine glands (pancreas, salivary glands), the liver, and the kidneys—can restore lost tissue by a combination of hypertrophy and hyperplasia.

Restoration of the liver after partial hepatectomy has been investigated most extensively. It is not certain that the Greeks knew about liver regeneration, but what must have been the first description of liver regeneration is contained in the legend of Prometheus by Aeschylus. Because Prometheus brought fire to the human species, the gods punished this innovator by tying him with chains to rocks in the Caucasus mountains. Every day a vulture came to eat portions of Prometheus' liver. This sadistic form of torture could not have lasted long if Prometheus' liver hadn't regenerated.

In the late 1800s, investigators excised 75% of the liver of rats and mice. It was observed that the remaining lobe doubled in size within 24 hours, and that the entire mass of the liver was soon restored.

New and old liver cannot be distinguished histologically, but the normal structure is never restored anatomically. Consequently, instead of speaking of liver regeneration, we should refer to this remarkable process as liver hyperplasia. But the term liver regeneration is so widely accepted that it would be futile to attempt to change it.

Since 1950 a great deal has been learned about the biochemistry of liver regeneration. During the early hours after partial hepatectomy, RNA polymerase increases, and new messenger, ribosomal, and transfer RNA is made. The cells hypertrophy and their content of cytoplasmic constituents doubles. Between 12 and 18 hours after partial hepatectomy, the genes coding for proteins involved in DNA synthesis become active. Thus, these genes which were repressed in the normal liver become derepressed after partial hepatectomy, and enzymes found only in small amounts, if at all, in normal liver appear in the regenerating liver. Among these enzymes are ribonucleotide reductase, which reduces ribonucleotides to yield deoxyribonucleotides: thymidylic kinase and other kinases that phosphorylate the mononucleotide to the triphosphate; DNA polymerase; and polynucleotide ligase. Histones and probably other proteins of unknown nature are also made in large amounts in preparation for the replication of DNA, which takes place between 24 and 30 hours after the operation.

Of course, the liver is not made of a single type of cell. In addition to the hepatic cells, there are mesenchymal cells (Kupffer's cells) with phagocytic properties, fibroblasts, and endothelial cells of the capillaries, veins, lymphatics, and arteries. To restore histological integrity these cells must not only proliferate, but they must reproduce exactly the social relationship that existed in the intact liver. Thus, after partial hepatectomy the hepatic, reticuloendothelial, and endothelial cells proliferate, but not necessarily at the same time. For example, the regeneration of the hepatic cells precedes that of the Kupffer's cells by approximately 24 hours.

Like wound healing or bone restoration, liver regeneration raises two important questions: what factors trigger the cellular hypertrophy followed by hyperplasia, and what factors regulate morphogenesis, or the social interaction among different types of cells? Two mechanisms have been implicated in triggering regeneration: the elimination of preexisting chalones and the elaboration of humoral factors that specifically stimulate liver regeneration. Neither the chalones nor the stimulatory factors have been extensively purified.

Practically nothing is known of the agents that stimulate morphogenesis.

It would seem that the ability of the liver to restore its lost substances would be a relief to the consumer of large amounts of alcohol, who every day destroys some hepatic cells. Alcohol destroys discrete cellular foci in liver. The destroyed cells are replaced by new cells through hypertrophy and hyperplasia. But the histological structure of the liver is never integrally restored, and foci of regenerating tissue are surrounded by thick bands of connective tissue whose origin is debated. Some investigators believe that the condensation of connective tissue results from the collapse of the liver fibrous stroma after the death of the hepatic cells, although others believe that it appears as a consequence of proliferation of the fibroblasts with collagen formation. In any event, the histological distortion is responsible for deviating the path of the blood vessel and causing a number of serious symptoms that do not need to be described here.

The combination of cellular death followed by hyperplasia of the hepatic cells and distortion of the connective tissue patterns is called cirrhosis. In typical alcoholic cirrhosis, the liver anatomy is changed and the surface of the liver becomes irregular and hobnailed. When the liver is sectioned, its consistency is found to be harder than that of normal liver, and examination of the surface of the section reveals foci of meaty tissue surrounded by thick interconnected layers of fibrous tissues. Cirrhosis of the liver is a condition in which many defense mechanisms (inflammation in response to cellular death, cellular hyperplasia, and proliferation of the connective tissues) are put into gear. However, the lack of proper coordination of the response to injury causes histological distortion, which itself becomes a cause of disease.

CAUSES OF DISEASE

INTRODUCTION

In spite of their heavy arsenal of defenses, humans are not eternal and they seldom die of old age. The aging process is a natural, time-dependent, progressive departure from optimal performance that inevitably afflicts all human beings who do not die from disease.

The fact is that most humans die from disease. It will be easier to understand what is meant by disease if we expand our definition of health. Health is a state of well-being, free of incipient disease, in which the human can perform without stress and without the use of drugs the physical and intellectual tasks for which he has been trained, taking, of course, the effects of age into account. Disease is a departure from health. The victim of asymptomatic and undetectable cancer of the pancreas is not healthy, although he may be unaware of this. In contrast, the loss of accommodation of the lens which affects the eyesight of every individual between the ages of 40 and 50 is not disease; yet the restriction of movement brought about by rheumatoid arthritis is a disease.

Diseases occur under different types of circumstances and include inherited defects and congenital anomalies, as well as those caused by conflict between the individual and the environment, defective nutrition, hormonal imbalance, electrolyte imbalance, overwhelming of the defense mechanisms, and excessive response to injury (the defense mechanism is out of phase with the intensity of the injury and therefore causes more damage than the injury itself).

HEREDITARY DISEASES

Some diseases are clearly hereditary and develop without obvious interference from the environment. We shall see, however, that it is quite likely that most hereditary diseases are caused by environmental injuries to one or both germ cells of the parents. The injury to the germ cell is then transferred from generation to generation by an ancestor born several centuries before the victim. Although it is almost certain that all hereditary diseases result from either alterations of the DNA or anomalies in the chromosomes, it cannot be excluded that in some cases primary injuries to the cytoplasm— for example, to mitochondrial DNA—could be inherited as well.

One can arbitrarily divide hereditary diseases into two major groups: inborn errors of metabolism, or diseases in which a single metabolic disorder is at the source of the symptoms, and an undefined group of diseases that result from complex chromosome anomalies, which in turn cause anomalies of development with anatomical and possibly metabolic distortion.

Some inborn errors may cause only mild symptoms that do not significantly affect the individual's well-being or life span. For instance, patients with fructosuria secrete large amounts of the sugar fructose in the urine without experiencing any discomfort. However, most inborn errors of metabolism (e.g., phenylketonuria and gout) cause severe disturbances of function, and the symptoms vary depending on the type of metabolic pathway that is altered. Only in rare cases, as in sickle cell anemia, is the link between the metabolic distortion and the symptoms known.

Inborn errors of metabolism result from alterations of genes, which are made of DNA. The DNA is part of the chromosome, but we do not know how genes are organized in chromosomes. In bacteria the expression of a single protein in the phenotype is regulated by a battery of genes (called operons), and genes with three types of different functions have been described: operator genes, regulator genes and structural genes. The operator gene is the switch that triggers the battery of genes. The regulator gene is a sort of rheostat that regulates the rate of synthesis of proteins coded for by the genes, and the structural gene is the actual blueprint of the protein.

Putting a gene in gear is analogous to starting a Xerox machine: the operator gene is the knob that triggers the replication of the gene battery; the regulator gene is the knob that can be set to determine the rate and number of copies to be made; and the structural gene is the blueprint for further copies.

The regulation of gene expression in the human chromosome is likely to be much more complex than in bacteria. In view of the large number of proteins and the complex mode of regulation at the level of the genome for their expression in the phenotype, one can expect inborn errors of metabolism to be numerous and to involve several different types of mutations in a single structural gene (for example, that of hemoglobin) and possibly alterations in regulatory genes. We shall return to this point when we discuss manifestations of injuries at the level of the units of specificity (Chapter 5).

In most cases, chromosomal anomalies of the germ cells result in distortion of fetal development and the growth of abnormal anatomic features with corresponding defects in function. A classic example, Down's syndrome, can be caused by several types of chromosomal anomalies. The most common type results from trisomy 21, in which the karyotype contains three instead of two chromosomes 21. The symptoms are unmistakable. In contrast, in the tetralogy of Fallot, a "congenital disease" of the heart, the type of chromosomal anomaly has not been identified.

CONGENITAL ANOMALIES

Congenital diseases are discovered at birth. They may or may not be hereditary. While the fetus grows it may be injured at an early stage by subtle environmental agents, such as alcohol or drugs, or by injuries to the placenta, often of unknown origin. Depending on the stage of development at which these injuries take place, various types of distortion occur. The distortions result from selective interference of the expression of the genotype in the phenotype which may be caused by selective interference with: (1) cellular proliferation or (2) the destruction of cells during embryonic development. An example of the former is thalidomide intoxication in pregnant women, which caused them to deliver children missing one or more limbs; an example of the latter is imperforation of the anus or the absence of separation of fingers. In humans the hand develops as an expansion of the embryonic bud of the arm, and early in their development fingers are held together by membranes. The separation of the fingers requires that the cells of these membranes die in a controlled fashion.

The development of the embryo involves not only cellular growth and differentiation, but also massive cellular destruction. This development is rigidly programmed at least in part by chemical agents.

The chemicals controlling embryonic growth, cellular maturation, and possibly selective death are called inducers. Few areas of research in biology have been more elusive. This is not the place to relate tales that come close to science fiction. Only present knowledge will be summarized. The intracellular mechanisms that direct cell differentiation during embryological development are highly susceptible to environmental changes. Alterations of ionic strength and ionic composition, shifts in CO_2 concentration, and addition of the proper concentration of ATP, steroids, nucleotides, fatty acid, and glycogen are all able to induce the formation of nerve tissues. Yet even in face of these discouraging data (because of the lack of specificity), Yamada and Tiedman continued to look for specific inducers (see Brachet). Available evidence suggests that at least some inducers are proteins, and a few have been identified: (1) a protein that converts gastrula ectoblast into mesoblast, (2) an inhibitor of the conversion of ectoblast into mesoblast, and (3) a protein that transforms ectoblastic into neuroblastic tissue.

Obviously, knowledge is still too primitive to come to definite conclusions as to the chemical structure and the exact role of inducers in the development of the embryo. But the time may be ripe for further investigations. Although there is at present no evidence for such speculation, it cannot be excluded that some congenital anomalies are caused by defects of production of inducers. Such defects could include overproduction, absence of production, or premature or belated production.

Less subtle causes of congenital anomalies are traumas resulting in damage to the placenta, the fetal organ that establishes contact between mother and fetus. In addition to many other functions, the placenta secures the transfer of nutrients from mother to fetus. If a portion of the placenta is destroyed, that portion of the fetus which depends on the damaged segment of placenta for survival stops growing, or if the fetal area is grown, it may degenerate causing a birth defect. Other factors that cause birth defects are fetal malposition, intrauterine crowding (e.g., presense of benign or malignant tumors of the uterus), and increased intrauterine hydrostatic pressure.

A classic example of a congenital anomaly of the heart is tetralogy of Fallot (Fig. 4-1) which is characterized by a combination of four anomalies: stenosis of the pulmonary artery, ventricular septal defect, right ventricular hypertrophy, and dextroposition of the aorta.

CONFLICT WITH THE ENVIRONMENT

Conflicts with the environment involve collision with physical agents, intake of toxic chemicals, and infection by live organisms.

Trauma

Trauma caused by blows, falls, or gun wounds results from the direct impact of parts or all of the body with solid masses. Such impact usually results in massive cellular destruction at the site of the impact. The skin or other protective epithelia, the underlying connective tissue, and even parenchymal organs such as liver and spleen can be damaged. When a limb is traumatized the bone may

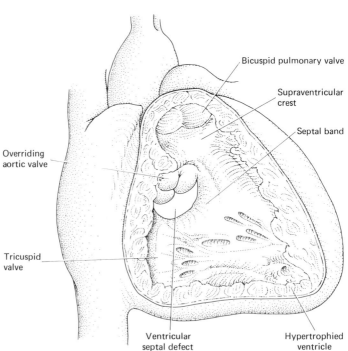

FIGURE 4-1. Tetralogy of Fallot.

be fractured. In trauma of the left quadrant of the abdomen, the spleen and sometimes the pancreas can be damaged.

A fall or a blow on the back of the head results in lesions of the frontal and temporal lobes. Similarly, impact on the right occipital lobe results in severe lesions in the left frontal and temporal lobes. A fall on the left temporal lobe results in greatest damage in the right temporal lobe. This paradoxical anatomical location of the injury is known as the contrecoup effect. There are many theories about the mechanism of the contrecoup effect, but none is entirely satisfactory.

The result of trauma is discontinuity of the covering epithelium, rupture of the blood vessels with hemorrhage, and cellular death with possible infections. The organism responds by the formation of clots, inflammation, and wound healing. An interesting form of reconstruction is that involved in the restoration of fractured bones (see Chapter 3).

More subtle forms of physical injuries are caused by: heat, resulting in burns of various degrees depending on the intensity of the heat; ultraviolet light, which causes sunburn; and ionizing radiations (X-, γ-, and cosmic rays), which at large doses administered to the whole body destroy proliferating cells and thereby deplete the bone marrow, spleen, and lymph nodes and interfere with epithelial proliferation. At low doses UV and ionizing radiations can cause mutations and cancer. We will return to these effects of radiation later.

Toxins

The intake of toxins is accidental, addictive, or iatrogenic. Toxins may also be produced by bacteria. Accidental intake results from ingestion, inhalation, and occasionally inoculation of venom from insects or snakes. Numerous foods contain toxic compounds. Usually the toxins are ingested in small concentrations, and the body either excretes or detoxifies the toxin. Occasionally some foods—e.g., mushrooms—contain extremely active poisons such as phalloidin and α-amanitine, or muscarine, which kill the victim within a short time after ingestion. Food can be contaminated by parasites or fungi that thrive in animals or plants, chemicals sprayed on plants to protect them against various kinds of pests, or drugs administered to animals to accelerate their growth. A classic example of contamination of plants by parasites is ergotism. Ergot is a reddish fungus that contaminates rye. It contains ergotamine, a hallucinogenic drug, and has been responsible for collective hysteria in both Europe and possibly in parts of the New World. The witch hunting of Salem in 1619 has been attributed to the contamination of rye by ergot.

Aflatoxin, a toxin found in a fungus that contaminates rice stocks in the Orient, is believed to be responsible for the high incidence of cancer of the liver in that part of the world. The addition of diethylstilbestrol to cattle feed has generated concern because the hormone is believed to increase the incidence of some cancers (e.g., those of the breast and uterus). There is, however, no conclusive evidence that diethylstilbestrol is carcinogenic unless it is administered in very large doses; therefore, it seems unlikely that the amounts of hormone found in meat are deleterious.

We are all aware of how the use of some pesticides has contaminated many of our foods. For example, until the 1960s cranberries in New England were sprayed with acetylaminofluorene, a substance known to be a most effective carcinogen in animals.

Inhalation of toxins may be involuntary, such as in cities with dense smog, or deliberate—for example, by smoking cigarettes. Addictive drugs have become such a common news item that it is superfluous to discuss them in detail here. They include barbiturates, amphetamines, alkaloids (such as cocaine, morphine, and heroin), and, of course, the ubiquitous ethyl alcohol.

Physicians often have no choice but to use drugs for patients who are in danger of becoming mutilated for life or dying it not treated. Some drugs are innocuous—at least for most patients—as is the case for penicillin. Only a rare person develops acute allergic reactions to penicillin. Aspirins, which are effective in relieving pain, have few or no side effects for most individuals unless they are blatantly overused. Then aspirins may cause hemorrhage of the gastrointestinal tract.

In other cases the physician and his patient are not as fortunate, and the drug administered may have rather unpleasant, and sometimes dangerous, side effects. For example, most drugs used against cancer have numerous side effects resulting from their ability to interfere with DNA synthesis and mitosis.

Biological Agents

This is not the place to recount the horrors of the dreadful epidemics of plague, typhus, cholera, typhoid fever, yellow fever, and smallpox, which wiped out tribes and armies, and sometimes entire cities. There is no doubt that at least part and possibly a great deal of history was shaped by such disease. At the beginning of his book *Doctors on Horseback,* Flexner remarked that more wars were lost by bacilli than in battle.

The contagiousness of plague and leprosy were known in the

Middle Ages, and appropriate measures were taken to prevent the spread of these communicable diseases, for example, by quarantining travelers coming from lands where the epidemic prevailed.* Yet most infectious diseases were believed to be caused by miasma in the air. Only after Pasteur and Koch laid the foundations of bacteriology were rational preventive measures introduced. Only between 1930 and 1950 were sulfonamides and antibiotics discovered.

The acquisition of food is central to the concept of infection. There may have been a time in the history of evolution when the first forms of animal life fed only on plants. Plants with the help of sunlight can convert atmospheric CO_2 and water and the nitrogen of the soil to essential building blocks, carbohydrates and amino acids. As the variety of the forms of life increased, some types of animals acquired their food by killing other animals and eating their flesh (predators), or by nestling in the body of the victim and consuming its substance (parasites).

Human beings may be the most effective predators, but they are also the victim of parasites.

Parasites are living organisms less evolved and smaller than the host, which in man live on the surface of the skin, the mucosae, or within the tissue, and feed on building blocks provided by their host. Who are these aggressors of what we think are the most highly evolved living beings on earth? Except for some worms, parasites cannot be seen individually and are therefore called microorganisms. Microorganisms include viruses, bacteria, protozoa, yeast, and fungi. Until the middle of the 20th century, bacteria were the undisputed rulers of the microscopic world. They killed millions in epidemics of plague, cholera, typhus, and leprosy. The agent of tuberculosis slowly consumed others by destroying their lungs, kidneys, intestines, adrenals, and even bones.

If the adventurous sailors of the Pinta, the Nina, and the Santa Maria who first saw the shores of the new continent did not return to Spain with spices and gold, it is generally believed that they harbored in their bodies a new type of microscopic aggressor, *Treponema* organisms, which cause syphilis. Syphilis spared no one, prince or pauper. It destroyed many a dynasty and mutilated millions in the large cities of Europe and the New World.

Microorganisms were first observed by a Dutch draper, who was also the caretaker of the Delft Town Hall, Anton van Leeuwenhoek (1632–1723). Anton ground lenses in his spare time and succeeded in building a primitive microscope. With his new instrument he

---

* Quarantine was instituted by the Venetians in 1403.

examined water and the scrapings of his teeth and discovered tiny moving organisms. He even noticed that the organisms seen in the scrapings of his teeth stopped moving if the material was collected immediately after he had drunk a cup of hot coffee. Yet another 200 years passed before Pasteur and Koch conclusively established that bacteria caused disease. This in spite of Fracastorius' suspicion that syphilis was caused by a living organism, and the demonstration of Francisco Redi in 1686 that "spontaneous generation"* of maggots in meat did not take place if the meat was carefully wrapped in paper and surrounded by a sheet of wire gauze.†

Bacteria Bacteria can be classified according to their mode of survival, general appearance, staining properties, the type of diseases they cause, and other characteristics. On the basis of mode of survival, one can distinguish two types of bacteria: the saprophyte, which lives in its own world feeding itself on dead plants and animals, and the parasite, which lives in symbiosis with the host. Parasites may be commensal and eat the same foods as the host without harming the host; other bacteria are pathogenic and cause disease and ultimately death.

Bacteria may be spherical (cocci), rod-shaped (bacilli), curved or comma-shaped (vibrios), or appear like spirals (spirilla) (Fig. 4-2). Cocci are distinguished by their relationship to each other and by their ability to pick up the gram stain. For example, staphylococci form irregular clusters, and streptococci form chains; both are gram positive. Diplococci are arranged in pairs and may be either gram positive or gram negative. Gaffkyae form a four-leaf clover in which all leaves are in one plane. Sarcinae are arranged in two superimposed layers of four (cloverleafs) and form a complex of eight separate bacilli (Fig. 4-2).

Bacilli are grouped according to their staining properties, their ability to form spores, their general shape, and their requirements for oxygen for growth.

In contrast to eukaryotic cells, in which the macromolecular constituents are organized to form identifiable intracellular structures (chromosomes, nuclear membrane, mitochondria, chloroplasts, endoplasmic reticulum, etc.), bacteria are in the prokaryotic group and lack specific intracellular organelles. There are no identifiable

---

* Aristotle claimed that living organisms emerged through spontaneous generation, and in *The Georgics* Virgil described a method for the spontaneous generation of insects. One can only conclude that Virgil was a great poet but a poor biologist.

† Pasteur conclusively disproved the theory of spontaneous generation with his memorable but simple experiments with swan neck flasks.

chromosomes, but a double-stranded DNA chain that may be circular carries the genetic information. There are no mitochondria or endoplasmic reticulum. Ribosomes are, however, found in bacteria. Other identifiable structures are the plasma membrane, made of lipoproteins (5–10 nm thick) that regulate the inflow of nutrients and the outflow of excretion products; in some cases the cell wall* (10–25 nm thick), a mucopeptide complex permeable to molecules less than 10,000 daltons; and the flagellum.† The flagellum (20 nm thick) is an organ of locomotion. It may be single and is then

---

* Mycoplasma are devoid of cell walls and are quite small (30–300 μm).

† Flagella should not be confused with fimbriae, which are numerous and smaller than flagella, and are not visible under the light microscope. Their function is not clear.

FIGURE 4-2.a. Types of bacteria.
1, Coccus; 2, bacillus; 3, vibrio; 4, spirillum; 5, spirochete; 6, gaffkyae; and 7, sarcinae.
b. Pneumococci (*top*) and streptococci (*bottom*) as seen under the microscope.

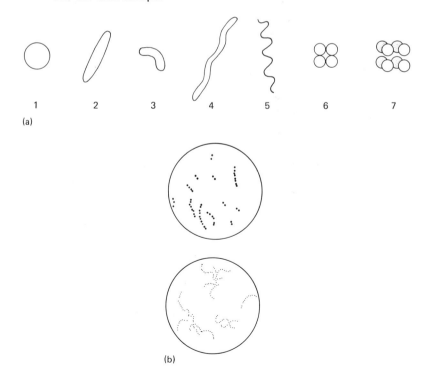

hooked at the stern of the bacterium like a long, flexible rudder. Some bacteria have multiple flagella distributed like many arms all around the cell's body. Flagella are made of a protein called flagellin, a relative of the muscle protein myosin.

The DNA of bacteria duplicates by simple replication. The bacterial cell duplicates by fission. This is in contrast to eukaryotic cells, which divide by the complex process of mitosis.

Bacteria sometimes contain inclusion bodies. These should not be confused with inclusion bodies found in viral infected cells. Bacterial inclusion bodies are believed to be nutrients. They are made of polysaccharides or lipids (open polymers of β-hydroxybutyric acid). Corynebacteria contain granules made of metaphosphates and nucleoproteins. The function of these granules is not known.

Sometimes the bacterial membranes form complexed invaginations called mesosomes seen in gram-positive and gram-negative bacteria. Although the role of mesosomes is not clear, three possibilities have been suggested. Mesosomes are the manifestation of cell division, the equivalent of eukaryotic mitochondria, or the site of attachment of DNA, thus facilitating DNA replication.

Some bacteria, for example clostridia, develop endospores. Spores are usually small, oval structures located at one extremity, in the middle, or at some intermediate position in the rodlike bacterium. They are surrounded by a thick membrane, the spore coat, which covers a laminar structure, the inner cortex. At the center of the spore is the core of cytoplasm, which contains a reserve of macromolecules needed for the supply of energy and for gene expression. Spores are remarkably resistant to environmental changes such as hyperosmolarity, the presence of detergent or toxic chemicals, heat, and radiation. For example, although regular bacteria are killed when heated at $60°–70°C$ for 10–15 minutes, only $100°–121°C$ will kill spores.

A list of pathogenic bacteria is shown in Table 4-1.

TABLE 4-1. Some pathogenic bacteria

Agent	Disease
I. Gram positive	
Mycobacterium	Tuberculosis
	Leprosy
Corynebacterium	Diphtheria
	Anthrax
Clostridium	Gas gangrene
	Tetanus
	Botulism

TABLE 4-1. (Continued)

Agent	Disease
Lactobacillius	Dental caries
Streptococcus	Tonsillitis
pyogenes	Impetigo
	Wound infections
	Otitis
	Scarlet fever
	Rheumatic fever
	Glomerulonephritis
faecalis	Urinary tract infections
	Wound infections
	Cholecystitits
viridans	Dental abscess
	Bacterial endocarditis
Staphylococcus	
albus	Usually not pathogenic
aureus	Boils
	Styes
	Wound infections
	Osteomyelitis
	Food poisoning
	Septicemia
II. Gram negative	
Neisseria	
meningitidis	Meningococcal septicemia
gonorrhoeae	Gonorrhea
Haemophilus	
influenzae	Bronchitis
	Pneumonia
	Meningitis (in children)
ducreyi	Chancroid
aegyptius	Conjunctivitis
pertussis	Pertussis
Brucella	
abortus	Brucellosis
melitensis	
Pasteurella	
pestis	Bubonic and pulmonic plague
multocida	Wound infection following the bite of an animal
Vibrio	
cholerae	Cholera
altor	

TABLE 4-1. (Continued)

Agent	Disease
Spirillum	Rat bite fever
Pseudomonas	Chronic otitis
	Urinary infections
	Wound infections
	Septicemia
Klebsiella	
aerogenes	Cystitis
	Pyelonephritis
	Purulent infections
pneumoniae	Pneumonia
Proteus	Wound and urinary tract infections
Salmonella	
typhosa	Typhoid fever
paratyphi	Paratyphoid fever
typhimurium	Food poisoning
enteritidis	Food poisoning
dublin	Food poisoning
Shigella	
dysenteriae	Bacillary dysentery
flexneri	
boydii	
sonnei	
III. Bacteroides	
Bacteroides fragilis	Wound infection
	Brain abscess
	Appendicitis
	Peritonitis
Fusobacterium	
fusiformis	Vincent's stomatitis
Borrelia	
vincentii	Vincent's stomatitis
recurrentis	European relapsing fever (louse borne)
duttonii	African relapsing fever
Treponema	
pallidum	Syphilis
pertenue	Yaws
carateum	Pinta
Leptospira	
icterohaemorrhagiae	Hemorrhagic jaundice (Weil's disease)
canicola	Canicola fever

Viruses A submicroscopic world in the biosphere is composed of a rich flora of organisms even smaller than bacteria. Indeed, these organisms—viruses—pass through the finest filters that retain bacteria. The diseases that viruses cause are certainly not a recent occurrence in the history of mankind. The great epidemics of smallpox and yellow fever were caused by viruses, as were the dreaded diseases rabies and poliomyelitis. Viruses have always caused numerous diseases in children: chickenpox, measles, German measles, mumps, etc. The affected child usually recovers fully and develops immunity to further infections. Rarely he or she develops complications such as meningitis (in measles) or orchitis (in mumps). Since vaccines have been developed for smallpox, yellow fever, and poliomyelitis, respiratory infections and viral hepatitis are the major diseases caused by viruses.

In recent decades virologists have transferred their concern from the traditional infectious viruses to the oncoviruses or viruses that cause cancer, at least in animals, and those viruses that cause slow viral disease.

As early as 1910 Peyton Rous demonstrated that a cancer that developed in the wings of some chickens could be transferred by an ultrafiltrate. Intact cells or bacteria were certainly not involved in transfer of the cancer. Between 1920 and 1940 Shope showed that a type of skin cancer in rabbits was also caused by a virus, and Bittner demonstrated that cancer of the breast in mice was at least in part due to the presence of a virus. Despite these investigations, viral carcinogenesis was considered to be a curiosity, and Rous' original work was almost ignored for 40 years until he was finally awarded the Nobel Prize in medicine.

In 1950 Gross opened the gates to the world of oncovirus by demonstrating that some mouse leukemias are caused by ultrafiltrable agents. Since then a rich fauna of oncoviruses has been discovered which causes a great variety of cancers in several types of animals.

There are too many oncoviruses to list here. Yet the only essential difference between oncoviruses and regular infectious viruses with respect to morphology or biochemistry is the special interaction between oncovirus and host. Moreover, although it is now easy to produce cancers in animals with the aid of oncoviruses, there is only circumstantial evidence that viruses influence the pathogenesis of cancer in humans. Among those cancers in humans that are suspected to be of viral origin are the melanomas, Hodgkin's disease, and leukemias.

Knowledge about the molecular composition of viruses grew with the development of viral oncology. Although the molecular structure of viruses is simpler than that of bacteria, various degrees of complexity exist even in the realm of viruses. A partial list of pathogenic viruses is given in Table 4-2.

As we have seen, viruses are composed of a core of nucleic acids and a protein coat called the capsid. The geometry of the virus varies with the type. Among the DNA viruses, adenoviruses, papovaviruses, and herpesviruses are isometric. In addition to the capsid, whose amino acid sequence is coded for by the virus, the herpesviruses have an outer envelope that is probably of host origin. Among the RNA viruses, the capsid is isometric in the picornaviruses, the reoviruses, and the arboviruses. It is helical in the orthomyxoviruses, the paramyxoviruses, and the rhabdomyxoviruses. Except for the picornaviruses, all the RNA viruses listed in Table 4-2 have an envelope.

The DNA and RNA oncoviruses are discussed in Chapter 5.

TABLE 4-2. Some pathogenic viruses

Agent	
DNA virus	
Adenoviruses	
Human type	Pharyngoconjunctival fever
	Keratoconjunctivitis
	Respiratory infection
Papovaviruses	
Papilloma virus	Human wart
Polyoma virus	Mouse tumors
Herpesvirus	
Herpes simplex virus	
Varicella virus	
Herpes B virus*	
Pox viruses	
Cytomegalovirus	
Variola virus	Smallpox
Vaccinia virus	
Molluscum virus	Molluscum contagiosum
RNA virus	
Picornaviruses	
Human	
Enterovirus	
Poliovirus	
Coxsackie virus	
(A and B)†	
Echo virus	
Rhinovirus	
Animal	
Foot and mouth virus	

TABLE 4-2. (Continued)

Agent
Arboviruses
A encephalitis
B yellow fever virus
Dengue fever virus
Louping ill virus
C encephalitis
Orthomyxoviruses
A influenza virus
B influenza virus
C influenza virus
Paramyxoviruses
Parainfluenza virus
Mumps virus
Measles virus
Rhabdovirus
Rabies virus
Oncovirus
(causes cancer only in animals)
e.g., Rous sarcoma virus, Bittner mammary cancer virus in mouse, etc.

* Causes a latent disease in monkeys but is lethal in humans.
† Type-A virus causes aseptic meningitis; type B causes myocarditis and pleurodynia.

Viruses that cause disease usually elicit cellular death and acute or chronic inflammation in the tissue where they lodge and proliferate. For example, poliovirus destroys the neurons of the anterior horns of the spinal chord and thereby interferes with the transmission of motor impulses to the limbs.

Slow virus diseases include a group of human and animal diseases—with an insidious onset and a slow but progressive fatal outcome—believed to be caused by viral agents. In some cases the viral origin is established (human subacute encephalitis); in others the virus has not been identified (kuru). Human subacute encephalitis is caused by the measles virus. There is, however, a long latent period between the measles infection and the panencephalitis. Kuru is a progressive encephalitis observed among the aborigines of New Guinea. The viral agent has not been identified, but human brain ultrafiltrates transmit the disease to chimpanzees. In New Guinea the disease was transmitted from one human to another because of the practice of cannibalism. Kuru is more prevalent among women because they are believed to eat more frequently than men the brain and other offal of their dead. But the pathogenic mechanism involved in slow virus disease is not known. The latency could be associated with a lysogenic state. Immunological interactions between host and virus have been invoked in the pathogenesis of disease caused by slow virus infections in animals.

Fungi Although some fungal infections, such as ringworm and athlete's foot, are common all over the world, other fungal infections occur only in restricted geographical areas. Coccidioidomycosis is endemic to the southwestern United States. Other types of mycoses that rarely cause serious disease in individuals not subjected to chemotherapy often cause death in cancer patients whose defense mechanisms have been annihilated by chemotherapy or combined chemotherapy and radiation.

In humans fungal infections are called mycoses. Depending on the degree of invasion of the human organism, one distinguishes between superficial and deep mycosis. Superficial mycosis affects skin, nails, hair, etc; deep mycosis is usually due to inhaled fungi and many spread from the lungs to other organs. Occasionally mycotic infections affect the mucosae of the genitourinary tract.

Among living organisms, fungi—like algae, protozoa, and slime molds—are categorized as Protista (Table 4-3).

TABLE 4-3. Four main categories of organisms

Nonera	Bacteria
	Blue greens
Protista	Algae
	Slime molds
	Fungi
	Protozoa
Metaphyta	Tracheophytes
	Bryophytes
Metazoa	Sponges
	Other animals

The phylum of fungi ranks fifth or sixth in number of described species. Fungi are found almost every place on earth, and in addition to playing an important role in ecology, many fungi contain extractable substances of economic and medical significance. Surprisingly, only 55 fungi are commonly pathogenic to human beings. Occasionally, saprophytic fungi become pathogenic in severely debilitated individuals.

The life cycle of a fungus includes a vegetative and a reproductive state. Fungi may be aquatic or terrestrial. Aquatic fungi are more primitive; their reproductive forms are round or oval cells that move with the aid of a flagellum. Terrestrial fungi are more developed.

The basic unit that composes the body of the fungus is a flexible, sometimes branched tubular structure called a hypha. The hyphae may grow extensively (Fig. 4-3). Numerous hyphae may form an intermeshed, irregular, reticular structure called the mycelium, or they may be more rigorously arranged to form recognizable structures such as mushrooms. The fungi are not made of true cells during the vegetative state. Instead they are formed of a mass of multinucleated cytoplasm. The cytoplasm is not divided by cell membranes, although incomplete septa have been seen in some types of fungi. True cells appear only during the reproductive state when a membrane surrounds a segment of cytoplasm containing a single nucleus.

The four major classes of fungi are Phycomycetes (nonseptate fungi), Ascomycetes (sac fungi), Basidiomycetes (club fungi) and the *Fungi imperfecti* (Table 4-4).

A complete description of all diseases caused by fungi is out of our reach. Only a few examples will be given. Ringworm fungi may infect the skin of the scalp or that of the rest of the body. The disease is manifested by ring-like lesions that are light gray or brown in the center at the beginning but later turn reddish. The ring lesions have an active reddish and slightly elevated edge that expands into a broader circle.

FIGURE 4-3. *Actinomyces (top); Candida albicans (bottom).*

TABLE 4-4. Major categories of fungi

Phycomycetes	Basidiomycetes
Water molds	Rust
Downy mildews	Smuts
Blights	Bracket fungi
Bread molds	Mushrooms
Ascomycetes	Toadstools
Yeasts	Puffballs
Molds	Stink horns
Powdery mildews	Fungi imperfecti
Truffles	A collection of types with
Cup fungi	unknown or incompletely
	known reproductive
	cycles. They are believed
	to include some asexual
	stages of Ascomycetes and
	maybe of Basidiomycetes.

Candida albicans are relatively large gram-positive organisms. Even in a normal individual, they may be found on the surface of the gastrointestinal, the (female) genital, and the respiratory tracts. The saphrophytes cause infections only when the defenses of the host are severely weakened. Candida is a yeast-like fungus, and microscopic examination reveals typical rounded and budding yeast cells as well as large elongated cells. The elongated cells are not true hyphae and therefore are called pseudohyphae. Pseudohyphae are nonbranching filaments, and in contrast to true hyphae they reproduce not by yielding spores but by constriction and division of the filament.

Cryptococcosis or torulosis is caused by a yeast that reproduces only by budding without hyphae or pseudohyphae. The agent that causes the disease is *Cryptococcus neoformans*, a 5–15-μm ovoid structure surrounded by a thick capsule. The fungus, found in soil and pigeon dung, causes an acute or subacute infection involving the brain, meninges, and sometimes joints. The presence of the organism elicits a granulomatous reaction. Rarely does the disease become generalized and affect the lungs, but when it becomes systemic it may be confused with lymphoma.

Blastomycosis has been observed in both North and South America. North American blastomycosis is caused by a dimorphic fungus, that is, a fungus that grows in a yeast form (thick-walled, small, ovoid or spherical budding organism) in the host and yields mycelia only when incubated *in vitro*, or when grown in the soil. The disease causes a granulomatous reaction. North American blastomycosis may be restricted to the skin or may be systemic; when systemic it often infects the lungs and the bones.

The South American form of blastomycosis differs from the North American by the localization of the lesions. Although the lesions may spread to any tissue, the lips, mouth, lymph nodes of the neck, and the ileocecal portions of the intestine are infected with predilection.

Coccidioidomycosis is a self-limited respiratory disease that occurs endemically; for example, in the southwestern United States. It causes mild respiratory disease or sometimes a more severe influenza reaction. Occasionally it develops into a disseminated granulomatous disease. In some cases of coccidioidomycosis, hypersensitivity reactions causing a skin rash have been described.

Histoplasmosis is also caused by a dimorphic yeast. The disease causes a granulomatous reaction usually confined to the respiratory tract and sometimes confused with tuberculosis. Histoplasmosis is more frequent in the United States than in Europe. The disease may either regress spontaneously or become disseminated and affect the liver, spleen, lymph nodes, kidneys, and brain.

Sporotrichosis is caused by a saphrophytic dimorphic fungus found in soil, plants, and woods. When it enters the skin, usually through a small wound, the fungus causes a granulomatous reaction that may lead to ulceration of the skin.

Actinomycosis also causes a granulomatous reaction followed by ulceration. It is usually seen on the neck or the face, but it sometimes affects the lungs or gastrointestinal tract. Actinomycosis is caused by a fungus usually found in animals, and two principal species of fungus have been described: *A. israelii* and *A. bovis*.

The defenses of the human body are not so effective that they can protect themselves from organisms more obvious and less subtle than the microbe, the virus, and the fungi. Although it is difficult today to have an exact account of human infections by protozoa and even worms, it is certain that these infectious agents continue to plague many humans.

Protozoa The human organism is not only the host of bacteria and viruses, but it can be infected by even more complex unit cellular organisms such as protozoa or even true metazoa.

Protozoa are unicellular organisms like bacteria, but their structural arrangement is much closer to that of eukaryotic cells. The nucleus is separated from the cytoplasm by a specific nuclear membrane. The nuclear proteins are organized into chromosomes that divide in a process akin to mitosis. The cytoplasm contains mitochondria. Protozoa move either through the propelling action of flagella, or in a way similar to the Roman galleys, which traveled the seas with the help of many oars, except that in protozoa the oars (cilia) are attached to the cell membranes and form flexible, filamentous structures. Some protozoa have neither flagella nor cilia. They move by expanding their cytoplasm in the form of pseudopods—structures resembling arms and legs that allow the cells to crawl and find their way wherever they desire to go.

The intestine harbors many parasites, including protozoa and metazoa.

Protozoa are the most nearly animal-like Protista. These unicellular structures store their genetic information in chromosomes, divide by mitosis, produce their major energy in mitochondria, and are separated from the surrounding environment by a cell membrane. Protozoa nourish themselves by pinocytosis and phagocytosis. Some are referred to as herbivorous and feed on bacteria and microscopic algae. Others are carnivorous and eat small metazoa or other protozoa.

Underneath the cell membrane there are contractile neurofibrils and myofibrils which are likely to secure the cell's movement. There are also vacuoles that are believed to serve to excrete excess water drawn into the cell to maintain the cell's intracellular osmotic pressure.

Protozoa can protect themselves against such threats from the environment as hypoosmolarity, electric charges, toxins, and changes in pH or temperature. In fact, protozoa have been claimed to be the most successful form of unicellular life. It is not known how many types of protozoa exist on earth. Some have estimated that more than 100,000 forms exist; it is surprising that relatively few are pathogenic. Among them are those that affect the gastrointestinal and urinary tracts—Amoeba, Giardia, Balantidium, and Trichomonas—and those that invade blood and tissues—Trypanosoma, Leishmania, Plasmodium, and Toxoplasma.

The *Entamoeba histolytica* is an ameboid protozoon 10–40 μm in size that causes ulceration of the ascending and sigmoid colon and the cecum. The *Giardia intestinalis* is a flagellate that may be found in the duodenum and the bile ducts and causes diarrhea in children. The *Balantidium coli* is a large protozoon (150 × 120 μ) causing either catarrhal infection, or sometimes ulceration of the large intestine. *Trichomonas vaginalis* (Fig. 4–4) is a saprophyte of the vagina which, under still undefined conditions, may cause an inflammatory reaction with pruritis. Trypanosomes cause sleeping sickness. At least two species are responsible: *T. gambiense* and *T. rhodesiense*. The tsetse fly sucks the blood of one infected human and then transfers the disease to another victim through an insect bite. *Trypanosoma cruzi* causes Chagas' disease, the American trypanosomiasis; the parasite is transmitted by an insect of the Triatoma genus.

Leishmania are intracellular, oval protozoa $3 \times 2\ \mu$ equipped with a small flagellum. They live inside macrophages. The *Leishmania* pathogenic for humans are transmitted by sand flies (genus *Phlebotomus*.) *Leishmania donovani* cause kala-azar, dumdum fever, or black disease; it affects mainly the spleen, liver, bone marrow, and lymph nodes. *Leishmania tropica* cause cutaneous sores. *Leishmania braziliensis* cause American mucocutaneous leishmaniasis. The reasons for the selectivity in infection site of these various types of leishmanias that look so much alike are not clear.

Toxoplasma gondii is a pyriform protozoon approximately $5 \times 3\ \mu$ in size. It infects nerve tissue with predilection, especially the retina and the brain.

Malaria is caused by plasmodia; four types are pathogenic in humans: *Plasmodium falciparum*, *P. malariae*, *P. vivax*, and *P. ovale*. All four are transferred by mosquitoes, the Anopheles. For centuries the plasmodia causing malaria have infected more people than any other infectious agents. The falls of Greece and the Roman Empire have been attributed to plasmodial infection.

The protozoa that cause malaria have a life cycle too complicated to be described in detail here. When a human being is bitten by an anophele, the insect inoculates its victim with saliva that contains the tiny parasite. The parasite follows the path of the blood

FIGURE 4-4. 1, *Trichomonas vaginalis*; 2, *Entamoeba histolytica*; 3, *Trypanosoma cruzi*; 4, *Leishmania donovani*.

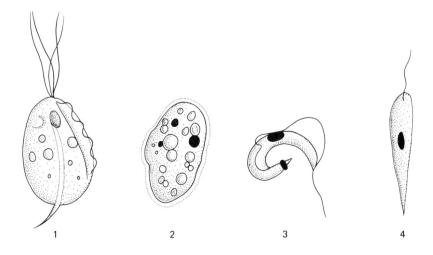

circulation, and reaches the liver cells where it multiplies until the cells burst. The progeny then return to the blood where they become engulfed in red blood cells. Red cell invasion occurs 12 days after inoculation. When the red cells are invaded, the victim develops severe chills and fever that may last for 2 or 3 days. The life cycle of *P. falciparum* is shown in Figure 4-5.

The victim's immune system is triggered, and antibodies against the parasite appear. This does not seem to unduly disturb the plasmodia, which simply assume new forms that are sexually differentiated. The male and the female are sucked in by the blood-thirsty Anopheles, and they mate in the belly of the host. Their progeny are transferred to the mosquito's salivary gland, and from there they infect other humans. This is a rather sketchy description of the complex life cycle of the malaria parasite. Yet it emphasizes how much we still need to learn about the interaction of the molecules of the host and those of the parasites if we hope to eradicate malaria by rational means.

Worms Many diseases in both the tropics and the temperate zones are caused by worms. There are two major groups of pathogenic worms: the roundworms or nematodes, and the flatworms or platyhelminths. The roundworms cause disease more frequently than the flatworms. Victims of worm infections usually suffer a chronic debilitating disease; occasionally, as in *Trichinella spiralis* infections, the victim may die suddenly. Flatworms are subdivided

FIGURE 4-5. Schematic representation of life cycle of *Plasmodium falciparum*.

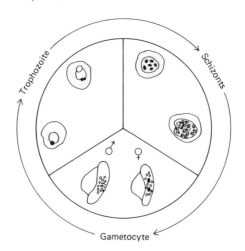

into flukes (trematodes) and tapeworms (cestodes). Although flatworms usually cause disease less frequently than roundworms, there are exceptions—namely, *Schistosoma* organisms, *Echinococcus granulosus*, and *Fasciolopsis* and *Clonorchis* organisms of the Far East. Worms cause a great variety of diseases; enteritis is probably the most frequent and hepatitis is second in frequency.

All worms have complex life cycles that often involve an intermediate host, which may be another mammal (sometimes a farm animal such as the pig, a house pet, e.g., a cat or dog, a fish, a snail, or a mosquito). The worm reaches man by four major routes: skin penetration, ingestion, inhalation, or introduction into the blood by insects.

One cause of worm infections is ingestion of undercooked meat. For example, eating undercooked pork may lead to infection by *Trichinella spiralis* and cause trichinosis. The parasite was discovered in 1835 by a medical student while he was performing his anatomical dissection. At first it was believed that *T. spiralis* traveled through the blood, but it is now suspected that the peritoneal cavity and the connective tissue are the preferred route for transfer of the worm from one organ to another. The adult worm does not really cause the disease. Larvae delivered by the female migrate to the skeletal muscle, particularly the diaphragm, intercostal muscle, larynx, and tongue. They become encysted, mature, and cause a chronic inflammatory reaction that ultimately leads to destruction of the muscle. If the affected muscle is the heart or a muscle essential to respiration, infection may cause death. *T. spiralis* are found not only in pigs but also in cats, dogs, wild bear, and other animals. The control of trichinosis is not very effective, and epidemics have been reported in the United States.

Another worm that invades humans as a result of eating undercooked meat is *Taenia solium*. The pig's muscle contains the cyst (*Cysticercus cellulosae*—the larval tapeworm), which is ingested by humans. The stomach juices digest the wall of the cyst. The head of the worm develops hooks that lodge it in the wall of the intestine. From the head emerges a long segmented tail, the unit component of which is the proglottid. The total length of the tail may be 70 cm, and each proglottid produces a large number of eggs that are shed in the feces. Uncooked beef contains a tapeworm called *Taenia saginata*, which has a life cycle similar to that of the *T. solium* except that the ox is the intermediate host.

Eating uncooked fish and crustaceans may lead to various kinds of infection by worms. Fresh-water fish harbor *Diphyllobothrium latum*, another tapeworm that lives in the human intestine.

In the life cycle of *Clonorchis sinensis*, a snail eats the egg, which develops into an intermediate form of development of the worm, the cercaria. When eaten by fresh-water fish, the worm develops further and becomes infectious to any man who eats the raw fish. The worm enters the common bile duct and from there lodges in smaller bile ducts and the liver. In the Far East a lung fluke, *Paragonimus westermani*, infects man after he eats raw crab or crab preserved in vinegar or wine.

Even eating of uncooked vegetable may be dangerous in some parts of the world. The intestinal fluke *Fasciolopsis buski* infects Far Easterners who eat raw water chestnuts and water caltrop. *Dracunculus medinensis* (or fiery serpent), common in parts of Asia and Africa, infects humans who drink unboiled water. The intermediate host is a fresh water creature of the cyclops species.

Worm infections caused by poor hygiene include hydatidosis, ascariasis, and trichuriasis.

Worms have a complex life cycle. In nematodes, apart from the egg, larval stages and the adult stage can be distinguished. The larvae differ from each other primarily by the size of the developing worm. Sexual differentiation usually becomes obvious at the fourth larval stage (Fig. 4-6).

The life cycle of platyhelminths is much more complex, and it differs in the fluke and the tapeworm. Let's consider the cycle of the *Clonorchis sinensis*. The adult fluke lives in the bile ducts of the ultimate host, man. The adults mate and the female lays the fertilized eggs, which pass in the human feces. Outside of the host the first larva develops within the envelope of the egg; it is a ciliated organism, the miracidium. Inside the snail the miracidia go through various stages of development: the sporocyst, which is free of cilia, produces buddings from its wall to yield the next larval stage, the radia. The radia are converted to tailed larvae, the cercariae, which emerge from the snail and are eaten by freshwater fish where they develop into metacercariae or young flukes. The fish are eaten by humans. Metacercariae move from the intestine to the bile duct, and the cycle is complete (Fig. 4-7).

Manson discovered the life cycle of microfilaria in the tropical zones of the Far East and presented his finding to the Royal Society. The reaction of the members of this venerable society is certainly not an illustration of its enlightenment. A member with a sense of humor reminiscent more of Voltaire's sarcasm than of traditional British humor asked Manson if the microfilaria carried a watch in their pockets. What may have been funny then is obviously stupid now. On the basis of her observation, Manson foresaw the cycle of malaria, which was established much later.

Interesting as the rhythmic cycle of parasites may be, few if any facts can explain it. What seems certain is that the microfilaria's rhythm is regulated by that of the host. What chemical message tells the parasite to move is not known. Changes in oxygen, glucose, and pH have all been invoked. Furthermore, all types of microfilaria migrate from their safe shelter in the lungs to the dark seas of the bloodstream at the same time.

Loa loa, a filaria that in many ways resembles the other filaria (same size and same shape), sleeps in the lungs during the night and swims in the blood during the day time because they must be sucked by daytime mosquitoes. Therefore, there seems to be little doubt that the clock is in the worm, but the host's metabolism decides what the migration cycle will be. The cycle of the parasite in the host is regulated by the life cycle of the preying mosquito. Thus, the parasite knows when the mosquito will come out, and it uses host chemical signals to move into the bloodstream.

The cycle is regulated by a response of the parasite to metabolic

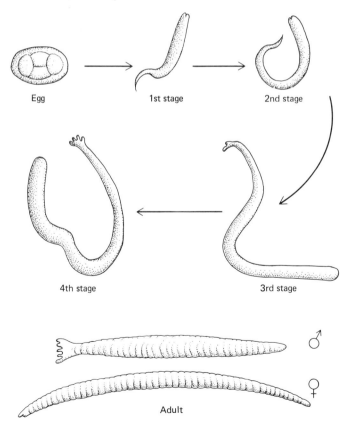

FIGURE 4-6. Schematic representation of the life cycle of the nematode *Ancylostoma duodenale*.

changes of the host. If the style of life of the host is changed, that of the parasite changes too. If a person infected with filaria sleeps during the day and works at night, the cycle of the microfilaria changes also. A list of some pathogenic worms is given in Table 4-5.

Arthropods Except for the butterfly, the ladybug, and a few other insects that have been romanticized because of the beauty of their colors or because of the imagination of poets, most insects are considered enemies of humanity and since antiquity men have tried to destroy them. The honeybee is a rare exception, so is the Spanish fly from which cantharidin (a venom that causes priapism and therefore was considered an aphrodisiac by the Romans and the Greeks) is extracted. The butterfly is not as innocent as it appears in its most glorious hours. At the larval stage the caterpillar

FIGURE 4-7. Life cycle of *Clonorchis sinensis*.

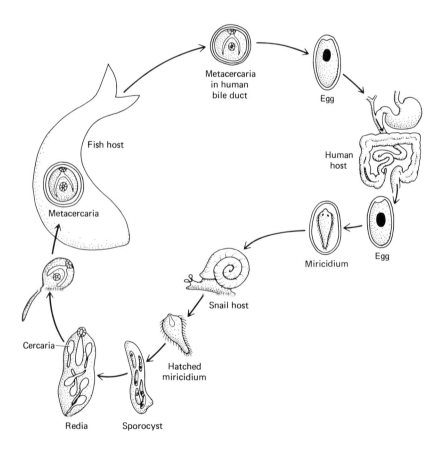

TABLE 4-5. Pathological worms

Mode of transmission	Host	Disease	Other
Skin penetration			
Nematodes			
Strongyloides stercoralis	Human	Enteritis	Hepatitis
Ancylostoma duodenale	Human	Enteritis	
Necator americanus	Human	Enteritis	
Ancylostoma braziliense	Human, cat, dog		Cutaneous larva migrans
Ancylostoma caninum	Human, cat, dog		Cutaneous larva migrans
Uncinaria stenocephala			Cutaneous larva migrans
Platyhelminths			
Schistosoma japonicum	Human, snail	Enteritis	
Schistosoma mansoni	Human, snail	Enteritis	
Schistosoma haematobium	Human, snail		Cystitis
Ingestion of undercooked food			
Undercooked meat			
Nematode			
Trichinella spiralis	Human, pig, and other mammals		Trichinosis
Platyhelminths			
Taenia solium (pork)		Enteritis	
Taenia saginata (beef)		Enteritis	

TABLE 4-5. (Continued)

Mode of transmission	Host	Disease	Other
Undercooked fish			
Platyhelminths			
Clonorchis sinensis	Human, snail, fish, cat	Enteritis	
Opisthorchis felineus	Human, snail, fish, cat	Enteritis	
Heterophyes heterophyes	Human, fish, snail		
Diphyllobothrium latum	Fish, copepod, human		Vitamin B_{12} deficiency
Paragonimus westermani	Crab, human		Pulmonary cyst
Undercooked vegetables			
Fasciolidae			
Fasciolopsis buski	Pig, snail, human	Enteritis	
Fasciola hepatica	Ruminants, snail		
Insect bites			
Nematodes			
Wuchereria bancrofti	Human, mosquito		Elephantiasis
Brugia malayi	Human, mosquito		Elephantiasis
Loa loa	Human, chrysops		Swelling of eyes
Onchocerca volvulus	Human, blackfly		"River blindness"
Poor hygiene			
Nematodes			
Those found in drinking water			
Dracunculus medinensis	Human		

Those found in human feces		
Ascaris lumbricoides	Human	
Trichuris trichiura	Human	Enteritis
Platyhelminths		
Hymenolepis nana	Human, mouse, rat	Enteritis
Those found in animal feces		
Dog feces		
Toxocara canis	Human, dog, cat	Visceral infection
Echinococcus granulosus	Human, sheep, dog	Hepatic and pulmonary hydatid cyst
Cat feces		
Toxocara mystax	Human, cat, mouse, insect	Visceral infection
Fox, wolf feces		
Echinococcus multilocularis	Human, vole, fox, wolf, dog	Alveolar hydatid cyst

has nettling hair that may enter the skin, the mucosae, or the conjunctiva and cause painful inflammatory reactions.

Insects are part of the phylum Arthropoda. The arthropods are segmented invertebrates with jointed appendages (legs and arms) and a protective exoskeleton. In addition to being a plain nuisance, they also can cause disease by secreting venom or by feeding on the host and either rendering him sensitive to the content of their saliva or transferring agents of disease. The venoms are believed to be neurotoxins. Indeed, they are secreted by one form of insect to kill the other. In man they cause mild edema, focal necrosis, and occasionally death.

Filaria are threadlike worms found in the tropics (*Onchocerca volvolus* or *Wuchereria bancrofti*). The female adult worm has two long uterine tubules that constantly generate eggs. The male is much smaller than the female. Once he has found his mate, he twists his tail around it and inserts needle-like structures into the female's vulva. The adult form lives in the lymphatics, especially those of the groin and the spermatic chord. If they are numerous enough, the worms cause scarring of the lymphatics with obstruction and edema. As the disease progresses, the genitals, scrotum or vulva, and limbs swell and the condition is called elephantiasis (Fig. 4-8). Fortunately, the small male Filaria is not always terribly effective at finding a mate and therefore only 2 out of 3 victims of filariosis develop elephantiasis.

When the male and female mate the egg is converted into a young larva called a microfilarium. The microfilaria are 200 μm long and 4 μm wide. They swim in the human blood until they

FIGURE 4-8. Left: elephantiasis. Right: adult Filaria.

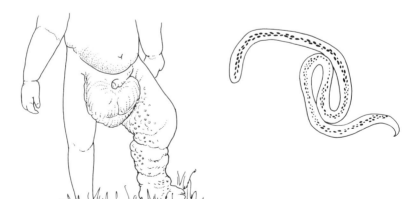

are picked up by a biting mosquito. In the mosquito they grow to 1.11 mm and become infective larvae that are transferred to other humans through mosquito bites.

The fascinating aspect of the life of the microfilaria is their rhythmic cycle. They migrate from the lung capillaries to blood vessels in a 24-hour cycle. For example, during the night *Loa loa* rest in the capillaries of the lungs; during the daytime they float in the bloodstream. The cycle is guided by their desire to be sucked by a mosquito which bites its human victims only by day.

In 3000 B.C. death caused by Hymenoptera was reported. A papyrus of 1500 B.C. contained a formula for an insect repellent.

The Bishop of Arethura, Saint Marcus, was martyrized in a most unusual form. He was placed in a basket, rubbed with honey, and left to be eaten alive by wasps.

Everyone has been the victim of an insect. Fortunately, most bites are relatively harmless and cause only occasional discomfort, but in some cases the insect is the host of viruses or worms and may cause severe diseases. In fact, probably many diseases transferred by insects are still unknown.

Recently Yale investigators have assembled data suggesting that a form of arthritis is transferred by a tick. A form of lymphoma (Burkitt's lymphoma) prevalent in some parts of Africa is, on the basis of circumstantial evidence, attributed to inoculation of the victim by a virus carried by an insect.

Insects can also cause disease directly by the venom they instill in the host. It may come as a surprise to many that the honeybee, *Apis mellifera*, kills more people in the United States than snakes. The bee has two glands: one secretes an acid, the other an alkaline product. The bee introduces a barbed stinger into the victim's skin, and special muscles pump the venom from a reservoir, the poison sac. The bee's venom contains a protein melittin, a hyaluronidase, an inhibitor of dehydrogenases, histamine-like substances, and activators of thrombokinase. When this venom is injected into a fly, it kills the fly. In humans the venom usually causes a minor skin lesion, but occasionally a victim is hypersensitive and develops a severe allergic reaction, or may develop hives, asthma, diarrhea, and even encephalitis.

Beekeepers are, of course, particularly at risk. When the bees of a hive are threatened, they secrete a substance with a special odor. This is a form of chemical message, a signal to attack. Each bee prepares itself for the attack and will commit suicide to save the hive. The bee starts by lubricating its stinger. Then it plunges it in the victim's skin with the venom reservoir; the muscles that empty the reservoir contract for at least 20 minutes, pumping more venom into the victim. After losing its stinger, the attacker bleeds to death. Although some individuals have survived the attack of a swarm, it is generally estimated that the venom of 500 bees will kill the victim.

Picnics, except those held in that portion of the biosphere that is free of insects, are often marred by attacks of yellow jackets, hornets, paper wasps, and ants. The wasp does not usually leave its stinger in the host. Its venom contains a number of chemical messengers causing inflammation (bradykinin, serotonin, histamine, and 5-hydroxytryptamine), acetylcholine, and a stimulant of smooth muscles. Reactions to wasp stings are often much more severe than those to bee stings.

Ants are found in most parts of the world, especially in tropical and subtropical regions. Some species are a health hazard because they contaminate food, cause irritating stings, or are vectors of disease. The life history of ants is one of the most fascinating chapters of biology, but it can be summarized only briefly here.

During the mating season winged ants, male and female majestically referred to as kings and queens, leave their nests, swarm, and mate. After the king has fulfilled his matrimonial duties, he is left to die. The young queen is responsible for securing the progeny. She usually does not return to the nest but finds a suitable spot to lay her eggs. The eggs hatch into small, worm-like larvae, the larvae grow into a quiescent form, the pupae. In some cases the pupae are surrounded by a silky cocoon. The pupae develop into adults, which are of three types: the winged kings and queens and the workers. Sometimes the workers are divided in various subclasses. The queen ant usually takes care of the first generation of larvae and pupae, but later generations are nursed by the workers. Satisfied with her accomplishments, the queen drops her wings and spends the remaining 15 years of her life laying eggs. Some of the injuries caused by ants are summarized in Table 4-6.

A classic example of a disease caused by mites is scabies. The mite, an obligate parasite, is transmitted by direct contact and lives within the keratinized layers of the epidermis. The females burrow into the skin and lay eggs as they travel. The eggs hatch into larvae that develop into nymphs, and the nymphs evolve into adult mites. Usual sites of infections are the fingers, back of the hands, elbows, armpits, groin, breasts, umbilicus, and penis. The activity of the mite is greatest when the body is warm, and therefore contamination is believed to occur during dancing or petting. It has been estimated that in World War I almost 225,000 combat days were lost in the United States Army because of scabies.

There are approximately 200 species of lice, but only two types infest man: *Pediculus humanus* and the *Phthirus pubis*. They are permanent parasites and leave the body as soon as it dies. It has

TABLE 4-6. Injuries caused by ants

Type of ant	Mode of attack	Venom	Injury
Ponerines (2-inch tropical ants)	Sting	Alkaloid	Welts, fever, paralysis swollen lymphatics, necrosis
Fire ants *Solenopsis saevissima*	Sting	Alkaline, contains amines	Hives, fever, asthma
Army ants *Eciton burchelli*	Sting	Cholinergic histamine-like substance	Aching pains, sweating
Desert ants *Pogonomyrmex barbatus*	Sting	Cholinergic histamine-like substance	Aching pains, sweating
Red bull ants *Myrmecia gulosa*	Sting	Hyaluronidase, kinins, histamine, melittin?	Welts
Formicinae	Bite	Formic acid	

been said that when Becket was murdered in the Canterbury Cathedral swarms of lice escaped the clothes that covered his dead body. Because they are permanent parasites they are transferred from one human host to another by direct contact. Louse infestation causes pruritus, which leads the victim to scratch, and scratching is followed by secondary infections (Table 4-7).

TABLE 4-7. Lice infesting humans

Pediculus humanus	Live in clothing and the body
Pediculus humanus capitis	Live in scalp hair
Phthirus pubis	Live in pubic hair, axillary hair (moustache, eyelashes)

One of the most disturbing consequences of infestations by arthropods is that they are frequently the vectors of viruses, bacteria, rickettsiae, and protozoa that may cause disease. Table 4-8 give some examples of diseases transmitted by arthropods.

This discussion on the relationship between parasites (whether they be bacteria, viruses, fungi, protozoa, or worms) and the host reminds us that a great proportion of humanity continues to suffer from diseases other than those (e.g., atherosclerosis and cancer) that traditionally plague technologically sophisticated societies. Yet we know little (except in the case of some bacteria and viruses) of the molecular composition of the parasite and the combination of molecular requirements that make a given parasite select a specific host. The problem of the relationship between parasite and host is one of ecology. The components that determine the environment of free-living animals are weather, food, other animals and living organisms, and shelter, or a place to live.

There are two major differences between the free-living animal and the parasite: the free-living animal depends on the outside environment and usually does not disturb its ecological balance; the parasite lives inside the host and often causes discomfort, disease, or even death.

The major problem raised by parasitism is that of the molecular combination in a given host which best suits a given parasite. Why is man the only host for *Taenia solium*, and why can *Trichinella spiralis* live at the expense of a number of hosts? A combination of factors that favor the survival of the parasite must determine these issues.

Surely the parasite needs water to grow, possibly glucose to feed on, in some cases vitamins that the host can provide, salts to main-

TABLE 4-8. Diseases transmitted by arthropods

Arthropods causing venenation	Arachnida	Sting	Bite	Insecta	Venom
Scorpion	+	+			Neurotoxin may cause ascending paralysis with respiratory failure
Spiders					
Black widow	+		+		Neurotoxin venom damages nerve endings and causes ascending paralysis
Tarantula	+		+		
Hymenoptera					Various types of venom that may cause hypersensitivity in the host
Bees		+		+	
Wasps		+		+	
Hornets		+		+	
Ants		+			

TABLE 4-8. (Continued)

Arthropods causing invasion	Arachnida	Insecta	Type of parasitism	Mode of transmission	Type of injury
Mites					
Sarcoptes scabiei	+		Permanent	Direct contact	Penetrate skin and cause pruritus, dermatitis
Maggots		+ (larva)	Part free living	Mother flies deposit their egg, usually in human wounds	Live on human tissues in ulcerated region in skin, mouth, nose, and orbit

Some diseases transmitted by arthropods

Vector		Agent	Disease
Mosquitoes	Tribe Culicini	Virus	Encephalitis St. Louis Japanese B Eastern Western Equine
	Aedes aegypti	Virus	Dengue fever
		Virus	Yellow fever

CAUSES OF DISEASE

Vector	Organism	Disease
(sandfly)	Worm	Leishmaniases
	Rickettsia	Bartomellosis*
	Virus	Sandfly fever (incapacitating disease of short duration)
Body lice *Pediculus humanus*	Rickettsia	Typhus
	Spirochete	Toxemia with febrile episodes
	Borrelia recurrentis†	
	Rickettsia	Trench fever (febrile disease, never fatal)
Rodent fleas	Rat flea *Pasteurella pestis*	Bubonic plague
	Rickettsia	Typhus
Ticks	Rickettsia	Spotted fever
	Virus	Colorado tick fever
	Spirochete *Borrelia duttonii*	Recurrent fever
Mites	Rickettsia *R. tsutsugamushi*	Japanese river fever
	R. akari	Mild spotted fever

* Causes warty excrescence on the skin of a granulomatous nature.
† Toxemia with febrile episodes.

tain its electrolyte balance, and oxygen for respiration. Still all these components could be found in many different hosts: why must the parasite select one or two special hosts?

Two groups of factors are likely to determine the selectivity. Some hosts provide optimal combinations of molecules for the survival of the parasite during all the steps of its life cycle in the host. These steps include infection after destruction of the epithelia of the skin or the intestine, maintenance of the parasite in the host until it is ready to reproduce and yield new infective agents, and maintenance of the progeny and provision of a favorable condition for their maturation into adult parasite. As we have seen, the completion of the entire cycle requires periods spent in different hosts (human and snails, human and fish, etc.) and often a free-living stage as well. In passing from the free-living state to the parasite state, outer protective layers are frequently lost (e.g., protective shell of the egg, cyst walls, sheaths of the larvae of nematodes).

The host does not usually remain passive to the invasion by a parasite. His defense mechanisms—acute and chronic inflammation and immunological reaction—are triggered. To survive, mature, and proliferate, the parasite must be able to overcome the defense mechanisms of the host.

Infections by trypanosomes are maintained because the organism is capable of changing its antigenic structure at intervals during infection.

In conclusion, although we know a great deal about the morphology and the life cycle of parasites that plague humans, we know almost nothing of the molecular combination that favors such parasitism. Information about such a molecular arrangement could be extremely valuable in controlling infections by more complex parasites.

NUTRITIONAL DEFICIENCIES

Nutritional deficiencies are of two major types: massive deficiencies such as starvation and specialized forms of malnutrition.

Specialized nutritional defects include protein, vitamin, and mineral deficiencies. Diets rich in carbohydrates but poor in protein result in a disease called kwashiorkor. Although kwashiorkor is most frequently observed in the third world, it has been described among poor whites and blacks in the United States.

Vitamins are either cofactors of enzymes (e.g., thiamine) or carriers facilitating transport (e.g., vitamin D). They cannot be synthesized by the human body, although other organisms can synthesize them. For example, bacteria can synthesize vitamin B_{12} and riboflavin but humans cannot.

In the absence of vitamins, maturation and differentiation are impaired in those cells that depend on the vitamins for enzyme activity or absorption of calcium or other nutrients. Thus, in vitamin B_{12} deficiency the red cells do not mature to produce normal erythroblasts; instead, large red cells loaded with hemoglobin, megaloblasts, appear in the circulating blood. These megaloblasts are the same as those seen in the embryo.

In the absence of vitamin C, the conversion of proline to hydroxyproline in collagen is impaired. As a result, the new collagen that appears in a wound is weak and wound healing is retarded.

Vitamins are either liposoluble or water soluble. Water-soluble vitamins are usually not stored in large amounts, so if a deficiency occurs symptoms soon appear. Lipid-soluble vitamins are often stored in liver or other organs, and an individual with a deficient diet may show no sign of deficiency for months.

Vitamin defects can occur as a result of: (1) dietary deficiency, (2) variations in the ecology of the bacterial flora, in which species synthesizing the vitamin are eliminated or a new type of microorganism appears that can catabolize the vitamin, (3) malabsorption due to absence of a carrier or excessive loss in the feces, (4) the presence in the diet of antimetabolites or antivitamins, and (5) inability to convert the precursor into an active coenzyme.

In the Orient where polished rice constitutes a major component of the diet, thiamine, a vitamin found in the chaff, is deficient. Beriberi, a thiamine deficiency disease characterized by polyneuritis and cardiac enlargement and edema, is frequently observed there. Thiamine deficiency has been discovered in cats and even in humans as a result of appearance in the intestine of bacteria that elaborate an enzyme which breaks down the vitamins.

Pernicious anemia is most often the result of the absence of intrinsic factors, proteins elaborated by the cells of the stomach which bind to vitamin B_{12} and are indispensable for its absorption. When intestinal lesions are associated with loss of fat in the feces (steatorrhea), liposoluble vitamins such as vitamins D, K, and A are lost in the stools and vitamin deficiency develops. Occasionally foods contain substances that bind strongly to a vitamin and thereby interfere with its metabolic role. For example, egg white contains a substance called avidin which binds to the vitamin biotin. Nevertheless, only one case of biotin deficiency has been reported: in a human who ate 40 raw eggs a day.

Among the mineral deficiencies, iron deficiency is probably the most frequent. Iron deficiency is discussed in more detail in Chapter 6.

HORMONAL IMBALANCES

Hormones are chemical mediators elaborated in one organ and transferred to another where they exert their effects either by stimulating the secretion of other hormones or by modifying the metabolism of that organ.

Hormonal imbalances are of two major types: decreases and increases in hormone levels. Decreases are caused by hereditary defects in the elaboration of the hormone, congenital absence of the organ that elaborates the hormone, destruction of the secretory gland, or interference with hormone secretion through metaplasia. Congenital cretinism is a group of diseases in which one of several steps of thyroxine metabolism is interfered with. As a result the victim lacks thyroid hormones and has symptoms of myxedema, a special form of edema of the skin and mucosal tissue that may markedly deform the face and the entire body. In Simmonds' cachexia all endocrine functions of the anterior hypophysis—a center controlling the secretions of thyroid hormone, sex hormone, adrenocortical and growth hormones, and other hormones—cease because of massive necrosis of the gland. Similarly, in Addison's disease, tuberculosis may destroy both adrenal glands and result in a condition characterized by weakness, hypotension, and hyperpigmentation of the skin. During World War II the victims of concentration camps who were fed almost nothing other than cabbage developed hypothyroidism because cabbage contains a compound akin to thiouracil, an inhibitor of some steps in the metabolism of thyroid hormones.

Increases in hormonal activities are usually caused by spontaneous or indirect hyperplasia of the endocrine glands. For example, the excess of growth hormones which causes gigantism in the growing child and acromegaly in the adult usually results from hyperplasia of the eosinophilic cells of the hypophysis, the cells that secrete growth hormones. Similarly, pheochromocytomas, tumors that develop by proliferation of the cells of the adrenal medulla, cause hypersecretion of catecholamine, the metabolites of which appear in the urine. Pheochromocytomas are also associated with hypertension and flushing of the face. An unusual circumstance in which hormones are secreted in excess occurs when the cells of a normally nonendocrine organ, such as the lung, in the course of the gene distortion that occurs in cancer switch to metabolic pathways in hormone synthesis.

As will become clearer later, two manifestations of cancer are an increase in the cellular population and loss of cellular differentiation. Some forms of lung cancer develop at the expense of rather primitive types of cells whose origin in the lung is not certain, but the cancer cells have the following features: they are relatively small, spindle-shaped, and have dark irregular nuclei and scanty cytoplasm. Because these cells resemble oat seeds, this type of cancer is called oat cell cancer. These tumors are sometimes associated with elaboration of a variety of hormonal agents such as hyperglycemic, antidiuretic, and corticotropic agents.

An indirect form of hyperplasia occurs in secondary hyperparathyroidism. The parathyroids, small glands located in the neck, secrete parathormone. The hormone regulates calcium mobilization from bone, and its increase leads to bone decalcification and high levels of calcium in the blood. Patients with renal deficiency lose excessive amounts of calcium in the urine; as a result the parathyroids are stimulated to secrete more parathormone, the cells proliferate, and the gland becomes hyperplastic.

BODY FLUIDS AND ELECTROLYTE IMBALANCES

Remarks on Origin of Life

There is no way to know how life began on this earth except by deductions on the basis of evolution of life as it exists now. Generally, it is agreed that among their ancestors, mammals (including human beings) must list some aquatic living organisms (born in the sea). Three stages in the origin of life on earth can be distinguished: the origin of the planet, the origin of the chemicals essential to life, and the origin of the first cells.

Five to ten billion years ago our solar system is believed to have been a big ball of fire that rotated in the universe like a top thrown by an expert. The rapidly rotating mass was made of gas; hydrogen was the principal atom present, but heavier atoms were also present. As in a centrifuge, while the burning mass rotated, the hydrogen condensed in the center and the heavier atoms spread to the periphery. This heavier gas belt, which crowned the central hydrogen mass that had become the sun, splintered in small masses of gas clouds that moved away from the sun. The gas cooled and condensed and the first planets were born.

One of these planets was the earth. At the beginning it was probably made of a mass of hydrogen and other atoms. The lighter atoms—hydrogen, nitrogen, carbon—remained at the periphery; the heavier—iron, nickel, etc.—moved toward the center; and the middle-weight atoms, such as silicone and aluminum, formed a middle layer. As the earth cooled bonds formed between atoms, and molecules of all sorts were formed. Although we do not know exactly which molecules were formed, it is fair to guess that they included H_2O, CH_4, NH_3, CO_2, HCN, and H_2. Many of these compounds remained in the atmosphere surrounding the earth because they were too light to be retained by the gravitational force of the earth mass. Then the earth cooled, the gas liquified, and the liquid solidified, at least in the middle shell. At the center temperatures remained too high to allow solidification. The water in the atmosphere flooded the earth and formed rivers and seas. The seas soon became enriched in the salts of the earth as a result of erosion due to the tides and constant churning by the bursts of molten lava.

The seas are believed to be the cradle of life. The simple gas molecules that were in solutions (CO_2, NH_4, CH_4, HCN, etc.) collided, and, if at the proper time their embrace was blessed with the gift of energy quanta provided by ultraviolet, X-, or cosmic rays from the sun, they blended into larger organic molecules. Some organic molecules must also have originated in the atmosphere through the electric discharges of lightning.

When electricity is discharged in a flask containing water, methane, and ammonia, many amino acids, fatty acids, and other organic compounds are formed. Adenine, a base essential not only for the construction of nucleic acids, but also as the major source of chemical energy, ATP, was artificially synthesized by long-wave UV light from hydrogen, cyanide, and ammonia. It is believed that similar processes might have taken place on earth during the prebiotic era. Amino acids have also been produced by irradiation of gas mixtures with long-wavelength ultraviolet light. Some believe that such UV light represented the most useful energy source for biological organic synthesis.

Although we can make some reasonable assumptions on the origin of the building blocks of DNA, RNA, and ATP, we are still far from understanding the formation of molecules essential for life. For example, did proteins or nucleic acids appear first? Whatever the mechanism, the macromolecular complexes appeared, and despite the vicissitudes of the environment they survived. How did they come together to form macromolecular complexes that could replicate? Were viruses (a core of nucleic acids and proteins) the first form of life, or did they appear much later in the biosphere as a degenerated form of life?

In any event, somehow the first cell developed. This certainly required that the essential macromolecules be brought and held together. There are at least three requirements for survival: the cell must replicate, have a source of energy, and be able to prevent leakage of its components from its internal environment into the medium. Some have proposed that this blending of the components of life took place in a small hole in the sand. But if fatty acids existed, could they have joined to form a bubble that entrapped some of the sea water and macromolecules indispensable for life?

Obviously, one could raise many questions as to how the first cell (or cells) ultimately developed into the complex multicellular organism that is the human being. To answer these questions is not within the scope of this book. The event in evolution relevant to this discussion is that as organisms became more complex they continued to carry a portion of the sea within themselves.

Seventy-two percent of the lean body mass is made of water; two-thirds is inside the cells (intracellular), one-third is outside the cells (extracellular) (Fig. 4-9). The mass of fluid in the body is maintained by epithelia. Some are totally impermeable to the penetration of fluid but may allow partial excretion (e.g., the skin and the epithelium of the respiratory tract); others absorb fluid (the intestinal tract).

The body fluids are not made of water alone. They contain ions (Na^+, K^+, Ca^{2+}, Mg^{2+}, H^+, HCO_3^-, etc.), proteins (albumin and γ-globulin in plasma), nutrients (e.g., glucose), excretion products (e.g., urea), or chemical messengers (e.g., steroids) that must be transported from one section of the body to another.

Different body fluid compartments vary in composition. For example, because the principal cation is potassium inside the cell and sodium in the extracellular fluid, sodium can enter the cell and be excreted by it when needed. This is achieved through a complex enzymic structure located in the cell membrane: the sodium pump. In addition to potassium, the cell contains Mg^{2+}

FIGURE 4-9. Relative distribution of body water.

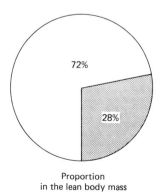

Proportion
in the lean body mass

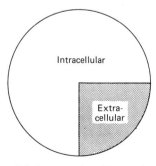

Relative proportion of intra-and
extracellular fluid

but little Ca^{2+}. The concentration of chloride is low in most the cells of the body except the cells of the gastric mucosa, the skin, and the kidney tubules.

Obviously the composition of the body fluids determines some of their fundamental physical properties, including osmotic pressure and pH. Despite the complexity of the compartmentalization of the body fluid and the passive and active exchange of ions from one compartment to another, the fundamental properties of the body fluid compartments are maintained remarkably constant.

Water and Sodium Balance

The maintenance of water and electrolyte balance is a complex process involving several organs (the kidney usually playing the principal role) and a number of hormones such as ADH, aldosterone, and the natriuretic hormone. A complete discussion of the electrolyte balance in the various body fluids is beyond the scope of this book. We shall restrict ourselves to a brief review of the factors regulating water balance and then list some conditions associated with electrolyte disturbances.

The maintenance of fluid volume is secured by balancing water loss and intake. Water is lost in the lungs, the intestine, and the kidneys, but the kidneys are primarily responsible for maintaining the water balance. The kidney increases water excretion in cases of excessive water intake and decreases water excretion by reabsorption under the influence of the antidiuretic hormone (ADH) in cases of excessive water loss (for example, through sweating or as a result of deficient intake).

The daily dietary sodium intake varies considerably, but most often it exceeds the sodium losses. Most of the sodium found in the plasma filters through the glomeruli, to maintain the sodium balance the kidney must therefore reabsorb sodium. Sodium reabsorption is regulated through at least three different mechanisms: (1) reabsorption through the proximal portion of the loop of Henle, (2) reabsorption in the distal portion under the influence of the hormone aldosterone, and (3) secretion of sodium in the distal portion of the nephron under the influence of a natriuretic factor. The details of the molecular mechanism involved in each step are too complex to be reviewed. Suffice it to point out that in the distal portion of the tubule aldosterone increases sodium excretion in exchange for potassium and hydrogen ions.

Accumulation of water in all (generalized) or part (localized—to a limb or body cavity) of the body is called edema. There are four major causes of edema: (1) increased hydrostatic pressure in the capillaries (e.g., generalized venous congestion in heart failure or thrombosis of the iliac veins); (2) decreased plasma colloid osmotic pressure (e.g., hypoproteinemia resulting from loss of protein in the urine in kidney disease, lack of synthesis of albumin in liver disease, or defective protein intake in malnutrition); (3) increased vascular permeability (primarily in cases of inflammation or hypersensitivity such as anaphylactic shock); and (4) sodium retention— when sodium is retained in the body fluids, to maintain osmolarity water is then retained as well, thus causing edema.

Sodium retention may be caused by reduced glomerular filtration, increased tubular reabsorption as a result of excess aldosterone secretion, or decreased aldosterone breakdown by the liver. Moreover, secretion of ADH contributes to water retention. Stimuli for aldosterone secretion are reduction of the intravascular portion of the extracellular fluid, sodium deprivation, and the formation of angiotensin. A reduction in the intravascular volume or a decrease in blood flow causes defective perfusion of the kidney that triggers the formation and secretion of renin and the conversion of angiotensinogen to angiotensin. Angiotensin stimulates the secretion of aldosterone. This pathogenic mechanism obtains in congestive heart failure, liver cirrhosis, and some kidney diseases including the nephrotic syndrome and glomerulonephritis.

One usually distinguishes between primary and secondary dehydration. Primary dehydration results from water depletion. Depletion may be caused by severe mental retardation with inability or refusal to take in fluids, or from the absence of water (castaways, desert travelers). As a result of water deficiency, the sodium concentration of the extracellular fluid increases and draws water from the cells into the extracellular fluid. This results in cellular dehydration, which causes thirst. Water depletion also stimulates the secretion of ADH, which promotes water reabsorption and ultimately causes oliguria.

Secondary dehydration is caused by loss of electrolyte-rich fluids: through diarrhea, vomiting, or as a result of pancreatic or biliary fistulas. Loss of electrolytes—in particular, sodium—results in hypotonicity of the extracellular fluid. The hypotonicity inhibits ADH secretion and water is lost in the urine to reestablish osmolarity.

Some other causes of electrolyte imbalances are presented in Table 4-9.

DEFECTIVE DEFENSE MECHANISMS

Some forms of disease result from defects in the defense mechanism. The defect may be hereditary and affect one or more molecules

TABLE 4-9. Causes of electrolyte imbalance

Type	Cause
Hypernatremia	Dehydration
	Excess water output
	Interference with ADH production
	Primary aldosteronism
Hyponatremia	Sodium depletion
	Overhydration
	Impaired water diuresis
	Congestive heart failure
	Cirrhosis of the liver
	Salt-losing syndromes
	Tuberculosis
	Bronchogenic cancer
	Idiopathic
Hyperkalemia*	Parenteral administration
	Ineffective excretion
	Chronic renal disease
Hypokalemia†	Inadequate intake
	Starvation
	Excessive loss
	Diarrhea
	Vomiting
	Excessive diuresis
	Adrenocortical hyperactivity
Hypochloremia	Vomiting
	Adrenal insufficiency
	Renal failure
Hyperchloremia	Water depletion
	Parenteral administration

* Hyperkalemia is associated with metabolic alkalosis, vacuolar degeneration of the tubular cells of the kidney, and necrosis of the myocardium with electrocardiographic alterations.

† Hypokalemia causes paresthesias, flaccid paralysis, electrocardiographic changes, and decreased blood pressure.

involved in defense, congenital and due to the lack of an anatomical structure needed for elaboration of the components involved in development of defense, or acquired and caused by destruction of the cells that elaborate one or more of the components responsible for defense mechanisms.

The lung contains a protein, the exact role of which is not clear, but which certainly inhibits trypsin, and therefore the protein is commonly referred to as α-antitrypsin. A hereditary defect in α-antitrypsin in a homozygous individual leads to chronic obstructive lung disease including bronchitis, asthma, emphysema, and bronchiectasia. A form of α-antitrypsin deficiency has also been observed in children, in whom it causes, by mechanisms unknown, necrosis of the liver cells with a vigorous, fibrous reaction and regeneration of the parenchymal cells leading to liver cirrhosis.

Classic hemophilia is caused by a defect in the female sex chromosome. Its history is too well-known to repeat here. The disease was transmitted by Queen Victoria to all courts of Europe and ultimately reached the Czarevitch of all the Russias, precipitating, if not causing, the fall of the Russian Empire. Because of the chromosomal defect, a protein indispensable for blood coagulation is not made. Consequently, the afflicted child is prone to develop hemorrhage. The slightest trauma may lead to uninterrupted bleeding under the skin and in the cavities of the abdomen, pleura, or joints. In the joints the blood remains locked in, distends the articular cavities, and causes severe pain.

Some pregnant women have marked bleeding tendencies because of a dietary deficiency of vitamin K. Vitamin K is needed for synthesis of another major component of blood coagulation, prothrombin, which is made in the liver. In severe liver disease inadequate amounts of prothrombin cause hemorrhage.

Diseases caused by the inability to mount adequate immune defenses are referred to as immunodeficiency diseases. The list of these diseases is too long to be discussed in detail here. Suffice it to point out that they may result from inadequate development. In *DiGeorge's* syndrome the third and fourth pharyngeal pouches fail to develop adequately, leading to markedly reduced cell-mediated immunity. Since the third and fourth pharyngeal pouches provide the anlagen for the thymus and parathyroid, the syndrome is associated with hypoparathyroidism and hypocalcemia. In another form of thymic hypoplasia caused probably by a defect in the development of the second and third pharyngeal pouches, the thymus is hypoplastic, cell-mediated immunity is defective, and antithyroglobulin antibodies develop which cause hypothyroidism.

Selected IgG, IgA, and IgM deficiencies have been described, as have combined deficiencies in IgE and IgM. These conditions naturally lead to an interference with the immune response. In IgA deficiencies, sprue-like symptoms develop. In IgG and IgM deficiencies, susceptibility to infection is often increased.

EXCESSIVE RESPONSE TO INJURY

More people probably are killed by the inadequacies of defense mechanisms than by any other means. We have discussed deficiencies in the defense mechanisms, but those are relatively rare compared to the untimely or excessive response to injuries by the defense mechanism.

It is well-known that many people die from heart attacks and strokes. Both conditions result from alterations of the arterial wall in the form of atherosclerosis, a complex vascular lesion which includes lipid deposition in the intima of the arterial walls, proliferation of smooth muscle cells, fibrosis, cellular necrosis, and formation of calcium soaps in the arterial wall. Quite naturally the lumen of the artery is narrowed, but the narrowing does not always cause serious damage to the heart or brain (unless the artery is occluded) because often collateral circulation develops. In most cases the formation of a blood clot, or thrombus, at the site of the arterial lesion causes death. The thrombus completely obstructs blood flow, and consequently the oxygen supply, to a large portion of the affected organ. Anoxemia is quickly followed by massive cellular death, and the necrotic tissue then elicits an inflammatory reaction which scavenges the dead cells in the heart. Attempts to repair the dead tissue may take place, not by proliferation of muscle, but by proliferation of connective tissue. In the brain the infarct may simply leave a hole filled with liquid, or it may occasionally be plugged by the proliferation of glial cells. An illustration of a spleen infarct is shown in Figure 4-10.

Once the vibrios of cholera have entered the lumen of the in-

FIGURE 4-10. Schematic representation of spleen infarct.

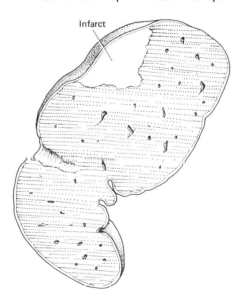

testine, they proliferate and elaborate a toxin. The toxin is not made by the bacteria but by a phage that contaminates it. The cholera toxin is irritating to the intestinal mucosa and causes a catarrhal inflammation that is so aggressive that large amounts of mucus and water are emitted through the intestine for the purpose of eliminating the toxin. This massive water diarrhea results in dehydration and electrolyte imbalances severe enough to cause death. In one of the massive epidemics in India, the simple administration of boiled saline saved 80% of the victims treated. It is interesting that the name "molecular disease" was first used by Snow in his famous book on cholera in which one of the chapters is entitled, "Cholera, a Molecular Disease."

In typhoid fever the bacillus penetrates the intestinal wall and proliferates in the Peyer patches. The Peyer patches react to the bacterial invasion by triggering massive proliferation of lymphoblasts and monocytes. Expansion of the Peyer patches distends the intestinal mucosa, which ultimately ulcerates leading to superinfection.

The overreactions of the defense mechanisms in thrombotic disease, cholera, and typhoid fever are overpowering. But sometimes the inadequacies of the defense mechanisms are much more subtle, as, for example, in immune reactions to injury.

Urine is made in the kidney. The unit structure of the kidney, the nephron, is like a twisted flower: its head is the glomerulus and its stalk is bent for physiological reasons that we will not discuss. The head is composed of entangled capillaries surrounded by an epithelial capsule resting on a basement membrane made of special connective tissue. The first step in urine formation is glomerular filtration. Blood passes through capillary bundles that are permeable to many blood components but not to blood cells or proteins with a high molecular weight.

Children infected by streptococci sometimes develop acute glomerulonephritis. The bacteria infect the glomerulus and other organs and elicit an immune reaction. The antigen that comes from the intruder complexes with the antibody and is deposited along the basal membrane causing thickening and interfering with renal function.

In other cases of glomerulonephritis, such as that accompanying Goodpasture's syndrome, the antibodies are not directed to any known intruder but for an unknown reason are aimed at one or more components of the basal membrane. It is likely that the immune system overreacts to many bacterial and viral diseases and thereby adds to the damage caused by the intruder.

Sometimes inappropriate reactions to injury occur at the metabolic level. In their eagerness to restore the genetic material, the enzymes that repair DNA may cause irreparable damage by forming double-strand breaks. The complex microsomal enzyme (mixed-function oxidase) that probably was devised by nature to defend the organism against toxins metabolizes some chemicals that would be harmless if they were not converted to active carcinogens.

CONCLUSION

The human body probably constitutes the optimal combination of atoms into molecules, molecules into macromolecules, macromolecules into subcellular structures, subcellular structures into cells, cells into organs, and organs into human bodies. Men are well prepared to defend themselves against injuries from the environment (UV and γ-irradiation, chemical toxins, trauma, infections) through metabolic restoration or inactivation, cellular and humoral immunity, inflammation, hypertrophy, hyperplasia, and wound healing.

When the regulation of normal metabolism or of defense mechanisms fails, humans become sick. Some diseases are present at birth (hereditary diseases and congenital anomalies), whereas others occur later as a result of physical, chemical, or biological environmental agents, nutritional deficiencies, hormonal and electrolyte imbalances, and defects in defense mechanisms.

INJURIES TO UNITS OF SPECIFICITY

INTRODUCTION

If organic function is to continue uninterrupted, new molecules must be identical to those that they replace. Therefore, for determination of specificity every cell contains a rigidly controlled mechanism which is ultimately enshrined in proteins.

Proteins interact with other macromolecules such as nucleic acids to modulate their template capacity, lipids or carbohydrates to provide the special selective properties of the membranes, other molecules to catalyze specific conversion of one type of compound into another, and antigens to form antigen-antibody complexes. Thus, proteins can arbitrarily be divided into various groups: structural, transport, messenger, catalytic, and immunological.

The structural proteins are part of the chromosomes or the cell membranes (periphery, mitochondria, or endoplasmic reticulum). The catalytic proteins are enzymes. They recognize specific molecules and catalyze their conversion into other molecules that are integrated within a cell structure, serve as substrates for other enzymes, are themselves enzymes with different properties, act as chemical mediators, or are eliminated.

Many primary injuries can be traced to molecules, and it is therefore, admittedly to a limited extent, possible to classify disease according to the molecular mechanism that causes most of the symptoms. For the sake of clarity, we shall consider molecular injuries as they occur in the five functional cellular units: (1) units of specificity, (2) catalytic units, (3) selective membrane units, (4) units of chemical mediation, and (5) antigen-antibody reactions. Although something is known about determination of specificity and enzyme function, much less is known about the mode of action of hormones, membranes, and antigen-antibody interaction.

The template unit is composed of the DNA template, the messenger RNA, the anticodon, the ribosome, and the polypeptide chain (see above).

PRIMARY INJURY TO DNA

Primary injury to DNA by radiation or chemicals may lead to strand breaks (single or double), base distortion, alterations in base sequences, and cross-links. Some of these lesions are repairable, others are not. In mammalian cells three major mechanisms have been discovered: (1) repair of single-strand breaks, (2) excision repair, and (3) postreplication repair. Let's consider some of the damage inflicted to DNA by various agents and then discuss repair and the limitations to repair.

X-Rays and some drugs (especially alkylating agents) cause, among other injuries, the development of breaks in one of the DNA strands. In mammalian cells most breaks probably are rapidly repaired, thus restoring the integrity of the strand. It is, however, conceivable that a few breaks in one strand are not immediately repaired, and double-strand breaks may develop as a result of a mechanism described below. At least two major kinds of breaks may occur. Some split the phosphodiester bond and yield an ending susceptible to binding by polynucleotide ligase, and others result in the rupture of the ring of the base or the sugar. Polynucleotide ligase repairs the strand break in the first instance; it requires a $3'$-PO_4 ending for activity. It is not known how the strand break is repaired in the second instance. An exonuclease could excise portions of the break, or the strand break could lead to a weakening of the hydrogen bond resulting in the separation of the two strands. The broken strand could then recoil and form new hydrogen bonds with a few bases of the unbroken strand. This process would result in the formation of a loop with a loose end. The portion of the loop connected to the remaining strand could be susceptible to an endonucleolytic attack followed by an exonucleolytic attack. Finally, the integrity of the double helix could ultimately be restored by the catalytic action of a DNA polymerase and a polynucleotide ligase.

Double-strand breaks can be caused by chemicals and by X-irradiation. Alkylating agents react with nucleophils adding an alkyl group. They may be monofunctional or polyfunctional and bind to many types of molecules, but it is generally believed that their major effect results from binding to DNA. The major site of binding is the N_7 of guanine. A monofunctional alkylating agent binds to guanine residues of one strand, whereas a bifunctional alkylating agent binds to guanine residues of opposite strands and causes cross-links. For reasons that are still obscure, the binding of alkylating agents causes double-strand breaks with a block of DNA synthesis.

X-Irradiation causes the formation of free radicals indirectly through water radiolysis and directly in macromolecules such as DNA. In some cases, at least after irradiation with large doses, double-strand breaks occur. The probability that a quantum of energy will hit the two strands at opposite sites is so low that other mechanisms for the formation of double-strand breaks must be considered. Such a mechanism will be discussed later. It is not known whether double-strand breaks are repaired at all in mammalian cells. However, an enzyme capable of repairing double-strand breaks in bacteria has been discovered.

What is the relevance of double-strand breaks to chromosome alterations? Some believe that the chromosome is essentially made of heavily supercoiled double-stranded DNA covered with proteins. If this is the case, at least some double-strand breaks may lead to chromosome breaks. Whether chromosome breaks are integrally repaired or not is not known, but what is certain is that chromosome breaks can be followed by intercalation, inversion, ring formation, or loss of genetic material (see below). Such changes are known to cause congenital anomalies if they are coupled with survival and cell reproduction.

DNA REPAIR

A distorted base is repaired in four steps: (1) incision of the strand close to the distorted base; (2) excision of the distorted base and several adjacent bases; (3) patching of the strand with new bases complementary to those of the other strand; and (4) restoration of the phosphodiester bond. The enzymes involved in the process are an endonuclease, which causes a break several steps removed from the damaged base; an exonuclease, which peels off the nicked segment including the abnormal base; a DNA polymerase, which restores the complementary strand; and a ligase, which inserts the repaired strand into the old strand by the formation of a phosphodiester bond (Fig. 5-1). If all the new bases inserted are complementary to that of the undamaged strand, then the repair is integral. But if a noncomplementary base is introduced, the repair is faulty.

There is overwhelming evidence that DNA repair occurs in mammalian cells, and we know something of the first enzyme involved in excision repair of DNA; namely, DNA endonuclease. The repair enzyme has a broad specificity and repairs damage caused by UV-irradiated, X-irradiated, and carcinogen-bound DNA.

FIGURE 5-1. Excision repair of UV-irradiated DNA.

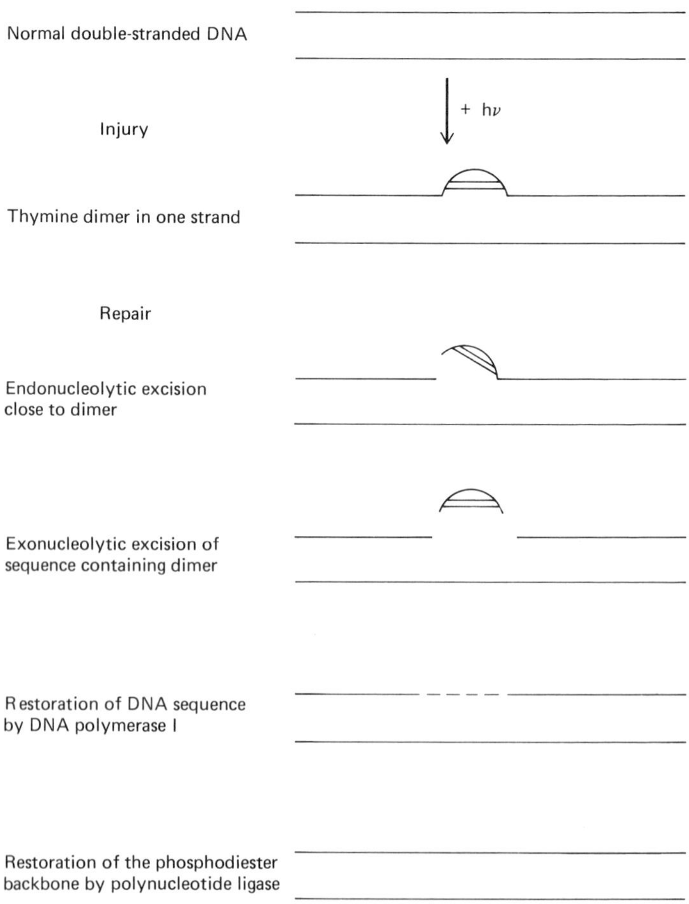

Normal double-stranded DNA

Injury
$+ h\nu$

Thymine dimer in one strand

Repair

Endonucleolytic excision close to dimer

Exonucleolytic excision of sequence containing dimer

Restoration of DNA sequence by DNA polymerase I

Restoration of the phosphodiester backbone by polynucleotide ligase

After UV irradiation of bacteria, not all thymine dimers are repaired by excision repair. Because part of the dimers disappear during DNA replication, this process is called postreplication repair. The exact molecular mechanism involved in postreplication repair is not known even in bacteria. The postulated mechanism is shown in Figure 5-2. Postreplication repair also occurs in mammalian cells, and it is believed to be much more prone to introducing the wrong bases into the repaired DNA strand than excision repair.

Restoration of base sequences was discovered in UV-irradiated bacteria. As pointed out already, UV irradiation of DNA induces,

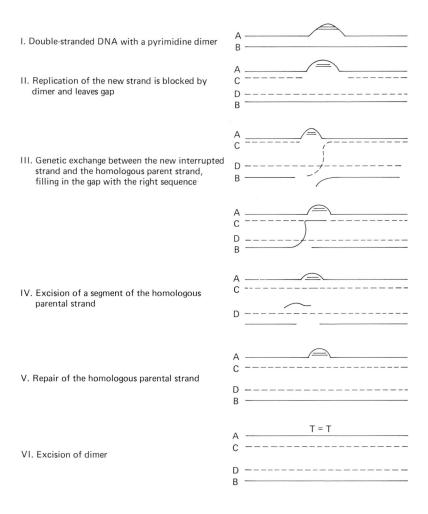

FIGURE 5-2. Recombination repair of DNA containing pyrimidine dimers (after Howard-Flanders).

I. Double-stranded DNA with a pyrimidine dimer

II. Replication of the new strand is blocked by dimer and leaves gap

III. Genetic exchange between the new interrupted strand and the homologous parent strand, filling in the gap with the right sequence

IV. Excision of a segment of the homologous parental strand

V. Repair of the homologous parental strand

VI. Excision of dimer

among other photochemical reactions, the formation of thymine dimers. Each time one sunbathes, some quanta of UV light reach the epithelial cells of the skin and possibly the underlying fibroblasts. The energy contained in the UV light is converted to chemical energy forming a covalent bond between two thymine residues. Remember that the DNA molecule is a neat, double-coiled helix, with each helix resembling a spiral staircase, the steps of which are spaced equally. Introducing a covalent bond between two thymine residues distorts the symmetry of the helix and a lump appears at the surface. To return to the analogy of the staircase, a similar distortion would occur if two of the stairs were tied together on one side causing them to tilt toward the outside. Such distorted DNA probably will not be adequately replicated unless the distortion is removed by one of the two mechanisms discussed above.

Carcinogens such as acetylaminofluorene bind to DNA and cause a similar distortion. They also must be removed if DNA is to be replicated successfully. It is fair to ask whether DNA repair is foolproof and complete.

It may well be that during excision repair the endonucleolytic attack is not always advantageous to DNA integrity. Experiments in which the effects of *in vivo* and *in vitro* irradiation of DNA were compared suggested that the endonuclease might add to the damage caused by radiation. It is not known how the enzyme affects X-irradiated DNA. The nicking probably is aimed at base distortion, but the nature of the base alterations has not been identified.

When DNA is extracted from the liver of irradiated animals and centrifuged in alkaline and neutral sucrose gradients, the most critical observation is the appearance of a dose-dependent peak at the low molecular weights in neutral sucrose gradients, suggesting the formation of double-strand breaks. In contrast, when DNA is irradiated *in vitro* and placed on sucrose gradients, only single-strand breaks are observed. However, if DNA X-irradiated *in vitro* is subjected to the action of the endonuclease and then placed on neutral sucrose gradient, a new peak appears, suggesting development of double-strand breaks. These findings can be interpreted in the following ways. X-Irradiation causes the appearance of single strand breaks and base alterations, both of which are potentially reparable: the single-strand breaks by the action of the ligase; the base alteration by excision repair. If the base alteration is located on one strand opposite to a single-strand break on the other strand, and the endonuclease reaches the altered base before the ligase has repaired the break on the opposite strand, then double-strand breaks develop (Fig. 5-3). Whether such double-strand breaks can be repaired in mammalian cells remains to be seen.

The recently proposed hypothesis that the induction process in cancer results from the formation of double-strand breaks with a high probability of error in repair emphasizes the significance of the presence of double-strand breaks. The restoration of the base sequence by excision or postreplication repair can lead to a mistake through insertion of the wrong base.

It is also possible that the break followed by excision of a fraction of a single-strand sequence could provide an opportunity for a viral sequence to be inserted between the loose ends of that strand. How such viral DNA may find complementarity in the other strands is not known. Again two possibilities need to be considered: (1) the virus is inserted and remains included in the strand and once included is no longer susceptible to enzymic attack, or (2) the presence of the virus in one of the strands leads to strand distortion, which becomes susceptible to either excision or postreplication repair. Again, such repair may be integral or faulty. Whether a break is followed by faulty repair or by incorporation of a total or partial viral sequence, the previous base sequence of the DNA has been altered, and when the new sequence is transcribed gene expression must be modified.

CONSEQUENCES OF INJURIES TO DNA

Clearly, such changes lead to mutations. If they occur in the germ cells and are compatible with cell survival, they are inherited mutations; if they occur in the somatic cells, they are somatic mutations. A mutated cell may be viable or nonviable. If the cell affected is nonviable, it dies and usually releases substances that result in an autoimmune or inflammatory reaction. A viable cell

FIGURE 5-3. Hypothetical mechanism for double-strand breaks after X-irradiation of DNA.

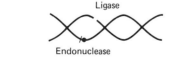

carrying a mutation divides, and the daughter cells carry new traits that may be either advantageous or disadvantageous to the species.

INHERITED MUTATIONS

There is evidence that after UV irradiation or the administration of carcinogens not all thymine dimers or all bound carcinogens are removed. The factors that restrict repair are unknown, but they are of considerable significance because the presence of altered bases probably interferes with the faithful replication of DNA.

Some base sequence distortions are transcribable into messenger RNA; others are not. If the distortions are not transcribed, the lesion will be fatal; if the lesion occurs in the portion of the DNA that must be transcribed for the maintenance of survival or cell reproduction, the dead cell may release substances that evoke inflammation and autoimmune reactions. In contrast, if the transcription of that portion of the altered DNA is not indispensable for survival but concerns, for example, side pathways such as the metabolism of an aromatic amino acid, a protein defect develops which is usually an inborn error of metabolism. However, an inborn error does not always result from a deletion during protein synthesis.

Some enzyme defects conceivably result from alteration in the amino acid sequence which may: (1) modify the catalytic site and make the enzyme unable to react with the substrate (differences in catalytic properties of glucose-6-phosphate dehydrogenase have been described); (2) render the enzyme exceedingly susceptible to proteolytic activity; or (3) modify allosteric sites, leaving the enzyme unsusceptible to normal regulation, such as activation or feedback inhibition. In some types of gout, the regulation of ribose pyrophosphorylase is suspected to be modified in such a fashion.

The primary alteration of the DNA may also result in transcription of a messenger RNA unusually susceptible to catabolic enzymes, thus preventing the translation of the protein in question. Such a mechanism is believed to explain some forms of thalassemia.

Theoretically, every protein of the body is susceptible to be translated erroneously as a result of damage to the DNA. When the amino acid sequence of the protein is altered, the initial damage must have affected a structural gene. This is certainly the case for sickle cell anemia.

In sickle cell anemia the base distortion in the DNA leads to an alteration of amino acid sequence of hemoglobin, which in some way distorts the cell structure by causing sickling. The obstruction of the capillaries by the sickle cell leads to anoxemia with extensive necrosis. Thus the molecular lesions inescapably climb the ladder of molecular, cellular, organic, and organismic levels of integration.

Many inborn errors of metabolism involve enzymes, and in most such diseases (except porphyria) enzyme activity is reduced. The defect could result from a lesion in the structural gene that interferes with its transcription, or from a regulatory gene lesion which prevents, by mechanisms unknown, the expression of the structural gene.

TABLE 5-1. Consequences of injury to germ cells: alteration of base sequence in DNA

Structural gene:	1. Block of protein synthesis (some enzyme defects)
	2. Synthesis of abnormal protein (e.g., sickle cell anemia)
	3. Synthesis of abnormal RNA susceptible to breakdown (some forms of thalassemia)
Regulatory gene:	Decrease in the rate of synthesis of the protein in question (some forms of thalassemia?)

SOMATIC MUTATIONS

What can be expected to happen if one or more base alterations take place in a somatic cell (any cell other than a spermatozoon or an ovum)? There are two possibilities: the cell may die or changes in the genome may be expressed in the cell's structure or function (the phenotype). If the mutation affects a gene responsible for the biosynthesis of a vital enzyme (e.g., an enzyme of the Krebs' cycle) or a vital structural protein (e.g., a membrane protein), the cell often dies. When only a few cells die a minor inflammatory reaction follows; when many die the inflammatory reaction is massive. It is also possible that as the cells die they release substances that are immunogenic and thereby elicit an autoimmune reaction (see below).

There are at least two possibile results of a mutation that is compatible with survival and is expressed in the phenotype. First, the function and appearance of the cell may be changed without affecting neighboring cells significantly. Perhaps some forms of metaplasia are of this type, although all forms of metaplasia are certainly not caused by base alterations in the DNA. Second, the mutation may involve the genes regulating the rate of cellular proliferation or the rate of cellular elimination. When the balance between cellular proliferation and cellular elimination is tilted in favor of cellular proliferation, a tumor develops. A tumor that simply proliferates, pushing aside normal tissue, is benign. Cancer arises when the tumor cell has undergone one or more mutations that provide it with, in addition to active proliferation, one or more survival advantages over the host cells. Such survival advantages include the ability to (1) use nutrients (amino acids, glucose, purine) preferentially, or (2) invade the surrounding tissue and colonize distant tissues by migration through lymphatic or blood vessels.

Thus, the benign tumor differs from the malignant one by the lack of invasion and metastasis in the former. A benign tumor can kill the patient only through compression of vital organs and not by exhaustion of the body resources.

Although we know a great deal about what causes cancer, we know little, if anything, about what primary molecular injury causes cancer. Known agents that cause cancer in animals include UV and ionizing irradiation, natural and synthetic chemicals, and viruses. The causes of cancer are not different in man. UV irradiation causes basocellular carcinoma of the skin. X-Irradiation, provided the dose is large enough and the latent period long enough (5 to 20 years), may cause cancer practically anywhere. Even relatively small doses of X-irradiation have been known to cause cancer in man. In the early days of diagnostic radiology when X-ray machines were not properly shielded, the incidence of leukemia among radiologists was ten times that of the general population. The incidence of leukemia is also believed to be higher among the survivors of the atom bomb than among the rest of the population in Japan. Similarly, children who were exposed to the atom bomb or to fallout in the Marshall Islands have developed more cancer of the thyroid and at an earlier age than the general population. In the first three decades of this century children were often treated with X-rays to reduce the size of the thymus. The incidence of cancer of the thyroid in these adults today is again greater than that among the rest of the population.

The first observation that chemicals cause cancer in humans was made by Sir Percivall Pott. He described cancer of the groin in English chimney sweepers which was caused by tar accumulation in the moist parts. Two hundred years later, it was shown that coal tar contains substances that are extremely effective in producing cancer in animals, for example, benzo(a)pyrene. Later it was observed that the weavers of Lancashire developed skin cancer of the thigh where oil constantly dripped from the weaving machine.

The development of the aniline industry coincided with the

appearance of a higher than normal incidence of cancer of the bladder among those who worked with those chemicals. β-Naphthylamine was found to be the culprit. Cancer of the liver in some Asiatic countries is believed to result from the contamination of rice by a fungus that secretes aflatoxin, which is carcinogenic in animals. Recently it was observed that a small population of workers in the plastics industry in which vinyl chloride is used develop a rare type of cancer of the liver called angiosarcoma. In some cases the chemical substances that cause cancer in humans are not known. For example, there is no doubt that heavy cigarette smoking causes lung cancer, but the exact chemical responsible for this is not known conclusively.

The effect of a chemical carcinogen can be enhanced by cocarcinogens, which are usually irritants. The role that cocarcinogens play in the incidence of human cancer is not known, and our understanding of their mechanism of action stems from animal experiments.

Shubik and Berenblum made a single application of a carcinogen to the skin of the mouse. Although the substance was an effective carcinogen, a single application did not cause cancer. Yet when the application was followed by several applications of oil extracted from croton seeds, the mice developed cancer. From these and many other experiments, the concept of two-stage carcinogenesis emerged, and a stage of irreversible initiation caused by the carcinogens was distinguished from reversible promotion caused by the cocarcinogens. The substance that acts as cocarcinogen in croton oil has been identified by Van Duuren.

Much remains to be learned about the molecular events in each stage of carcinogenesis in animals. Moreover, since it is impossible to experiment in man, it is not known how much a combination of carcinogens and irritants may contribute to the development of cancer in man. For example, is the carcinogenic effect of smoking due to a combination of deposition of the carcinogen and irritation of the bronchial epithelium by some other components in cigarettes? Does the irritant in smog contribute to the carcinogenic effect of cigarettes, or does the irritation of the bronchi caused by smoking hashish contribute to the development of lung cancer? One could also imagine that irritation of the intestinal epithelium by some laxatives taken repeatedly combined, for example, with a carcinogen excreted in the bile or produced by bacteria could cause cancer of the colon.

After contemplating these possibilities, which may be realities, one can understand the importance of discovering the molecular changes that take place in the stages of initiation and promotion.

What is known about the way chemicals cause cancer? Much

information on the metabolism of carcinogens is available. Let's take, for example, benzo(a)pyrene, the substance extracted from tar and suspected to be the carcinogen in cigarettes. Benzo(a)pyrene, is not by itself carcinogenic. It is converted to epoxides which then bind to DNA. Acetylaminofluorene (AAF), which was used as an insecticide and sprayed on cranberries, causes many types of cancer in animals. When fed to a rat it causes cancer of the liver. However, AAF is not the active carcinogen. A group of microsomal enzymes called mixed-function oxidases (because they hydroxylate drugs such as phenobarbital and many different chemical compounds) hydroxylate AAF, which is further converted to a sulfate ester and ultimately binds to DNA, proteins, and polysaccharides. Some other carcinogens, such as propionolactone, bind directly to DNA.

On the basis of these two examples, we can now outline a general pattern for the molecular conversion of a substance into an active carcinogen. After ingestion or inhalation, the environmental carcinogen binds directly to macromolecules or is converted by the cell's enzymes into an active carcinogen with electrophilic properties. The active carcinogen binds to nucleophilic macromolecules.

Although viruses are suspected to cause some cancers (lymphoma, cancer of the lymph nodes; leukemia, cancer of the white blood cells; and breast cancer) in humans, convincing evidence for this is not available. There is, however, no doubt that viruses cause cancer in animals.

Viruses, the smallest form of life, are primarily composed of nucleic acids (DNA or RNA) and proteins. In the simplest viruses the proteins are of two major types: coat proteins and enzymes. Both types of protein are coded for by viral nucleic acid, but the virus uses the molecular machinery of the host for protein synthesis (ribosome, transfer RNA, etc.).

To synthesize protein, viruses may cause various kinds of cell damage including death and conversion of a normal into a cancer cell. This conversion can be achieved in tissue culture, where it is called transformation. Many changes take place in the transformed cell: a large number involve the cell membrane, others affect the host's DNA. When normal fibroblasts grow in culture in a Petri dish, they form a single layer of cells. Transformed fibroblasts crawl over each other and form several cell layers arranged in a disorderly fashion. The orderly growth of the normal cell is believed to relate to some special properties of the cell membrane, which emits signals to the cell's genome and tells it to stop growing. This phenomenon is called contact inhibition of growth. It is believed that the cancer cell has lost this property because of alteration of its membrane. Although we do not know what molecular alterations are responsible for the loss of contact inhibition and how these membrane changes come about, several biochemical changes have been described in the membranes of transformed cells.

The presence of cancer viruses also induces important changes in the DNA of the host. The polyoma virus, which causes cancer in mice, is a DNA virus. During transformation, the DNA strands of the virus are inserted into the host DNA. Rous sarcoma virus, which causes cancer in chickens, mice, and monkeys, is an RNA virus. RNA cannot be introduced into DNA, but independent studies done in Temin's laboratory and in Baltimore's at the National Institutes of Health have shown that the viral RNA is converted to DNA by a special enzyme called reverse transcriptase. The viral DNA strands are then inserted into the host DNA.

With this preliminary information on the mode of conversion of a normal into a cancer cell, it is possible to raise some questions about the cancer cell.

In the cancer cell the normal expression of the genes into the special protein mosaic (structural, catalytic, etc.) that gives the cell its unique functional and morphological characteristics is distorted. Thus, cancer is a distortion of differentiation. The distortion in gene expression provides the cell with survival advantages over most, if not all, cells in the host. The cancer cell preys on the nutrients of the host, with no regard for histological and anatomical boundaries. A cancer that starts in the epithelium of the bladder invades the subepithelial layers and penetrates the perineal tissue and even the rectum.

Moreover, some cancer cells may come loose from the mother tumor; travel in the lymphatic or blood circulation; settle in lymph nodes, liver, lung, or bone marrow; and proliferate at sites distant from the primary tumor. This process is called metastasis. Obviously, the survival advantages of the cancer cell are transferred from one cell generation to another. The initiating event must therefore involve the genome directly or indirectly.

For the sake of argument, let's center our discussion on the conversion of a normal into a cancer cell in those cancers caused by chemicals and radiation. The chemicals bind to DNA, RNA, proteins, and polysaccharides. They could therefore cause distortions in many metabolic pathways and alter some important cellular structures, such as the cell membrane. Feedback loops between the damaged membrane and the genome could exist in which the damaged cell membrane would tell the genome to alter its gene expression, and the freshly instructed genome could in turn produce a new type of membrane. This is not inconceivable if one thinks of how the membrane of a new cell is made: by expansion of the old membrane after new DNA is synthesized and chromosomes have duplicated and separated.

We know little of the mechanism that controls gene expression in mammalian cells. Clearly, the genetic information is stored in DNA. However, only part of that information is used from one generation of cells to another in a given tissue. When a liver is subjected to regeneration after 12 successive partial hepatectomies, it is always hepatic cells that regenerate. The selection of the phenotype is transferred from generation after generation not only in hepatic cells but also in epithelial cells of the skin, fibroblasts, etc. Nonhistone proteins are believed to play a role in determining the expression of the phenotype. Their exact mode of action is not known, but it is not inconceivable that the carcinogen modifies the mechanism regulating expression of the genotype into the phenotype. This could occur by binding to nucleoproteins (histones and nonhistone proteins). Of course, this remains to be proven.

Indirect mechanisms of alterations of gene expression cannot be overlooked, but a great deal of evidence suggests that cancers caused by chemicals and radiation result from direct damage to the genome.

As we have seen, UV light, ionizing radiation, and chemical carcinogens damage DNA. In all three cases the damage is repaired, but probably only partially. When the lesion is not repaired, as is the case in xeroderma pigmentosum, a disease in which excision (as shown by Cleaver of the University of California, San Francisco) or postreplication repair is inhibited, the incidence of cancer is high. Although their skin may be pigmented, patients with xeroderma pigmentosum are extremely sensitive to ultraviolet light. Their skin appears old at an early age and they develop skin cancer, sometimes in their teens.

Thus, there is strong evidence that DNA is the primary target of the carcinogenic agent and that therefore the conversion of a normal into a cancer cell results from one or more somatic mutations. Logically, if the cancer cell results from somatic mutation, radiation and chemical carcinogens should be mutagenic. UV light and ionizing radiation have long been known to be mutagenic, but only recently was it clearly demonstrated (mainly thanks to the work of Ames) that under the proper conditions most known chemical carcinogens are mutagenic for bacteria and mammalian cells.

In conclusion, a plausible hypothesis for the molecular mechanism of conversion of a normal into a cancer cell is that the carcinogenic agent induces one or more somatic mutations. Could a single somatic mutation cause a distortion in gene expression that would provide the cancer cell with the survival advantages already described? Such a mutation could lead to the elaboration of a DNA polymerase susceptible to making mistakes in DNA replication. This would result in a cascade of mutations, and some of the cells might be expected to develop a combination of mutations that would turn them into cancer cells.

One could also think of another mechanism by which a single gene mutation could lead to multiple phenotypic changes. If, for example, a repressor were made of two polypeptide chains—one with a sequence common to all repressors, the other with a sequence specific for each repressor—then a lesion of the DNA sequence coding for the first of these polypeptide chains could alter various repressors, and the alteration would in some way depend on the status of gene expression in the injured cell. Such a mechanism of injury could lead to a variety of distortions of gene expression, a combination of which might provide survival advantages.

Whether cancers caused by viruses are the result of a somatic mutation resulting from the insertion of the virus in some critical portion of the host DNA remains to be seen. Other theories have been proposed. The dominant one, the oncogene theory, assumes that maybe a million years ago an RNA virus was converted to a DNA sequence that was incorporated into the host DNA forming a new gene, the oncogene. Normally the gene is repressed, but after exposure to UV and ionizing radiation and chemical carcinogens, the gene becomes derepressed and the normal cell is transformed into a cancer cell.

INTERFERENCE WITH DNA SYNTHESIS

Cells can be divided into three categories with respect to their ability to synthesize DNA and undergo mitosis. Some cells, such as neurons of the brain, never divide or undergo DNA synthesis. Others, such as cells of the bone marrow and those of the epithelia of the gastrointestinal tract, divide constantly. The cells of the liver, pancreas, and many other organs are between these two extremes in that their rate of division is low under normal conditions, but it rapidly increases (especially in the liver) after the proper stimulus has been administered. For example, excision of three-fourths of the liver stimulates the remaining portion to proliferate rapidly. In fact, the number of cells doubles within 24 hours, and the liver usually regains its normal weight within 2 days after partial hepatectomy.

DNA synthesis is obviously critical for certain organs and less important for others. Interference with DNA synthesis has long been shown to be associated with cellular death. Such interference may result from direct damage to the DNA molecule. We have already discussed some aspects of the damage caused by UV irradiation and X-irradiation. Interference with DNA synthesis may also be caused by inactivation of enzymes involved in the last steps of DNA synthesis, substrate deprivation, and interference with mitosis.

Enzymic Block

Examples of the block of DNA synthesis resulting from interference with the essential enzymes include block of nucleotide reductase by hydroxyurea and block of both nucleotide reductase and DNA polymerase by the antimetabolite cytosine arabinoside. When DNA synthesis is blocked in lymphocytes or in growing cells of the intestinal mucosa or bone marrow, extensive necrosis of these cells follows. However, if these drugs are administered to a partially hepatectomized animal, there is a delay in DNA synthesis in regenerating liver, but the liver recovers.

Substrate Depletion

Direct block of DNA synthesis by substrate depletion is not known to occur *in vivo* in mammals, but it occurs in bacteria that are dependent on thymine for survival. Exclusion of thymine from the culture medium blocks DNA synthesis and leads to cellular death.

Antimetabolites

A more subtle form of interference with DNA synthesis often used for therapeutic purposes in mammalian cells consists of administration of any of four types of antimetabolites: folic acid analogs, glutamic acid analogs, and purine and pyrimidine analogs.

Folic Acid Derivatives Folic acid is a vitamin indispensable for the biosynthesis of the building blocks of DNA and RNA and some amino acids. As is the case for most vitamins, folic acid is not the active compound, but one of its metabolites, tetrahydrofolic acid, is. Inasmuch as pernicious anemia is primarily due to a defect in folic acid and vitamin B_{12}, it was suspected that folic acid might help in the therapy for leukemia. However, when such therapy was attempted, it was discovered that the patient's condition worsened after folic acid administration. In contrast, administration of a folic acid–free diet slowed the course of the leukemia. Attempts were then made to prepare metabolic antagonists of folic acid. The first antagonist discovered was aminopterin, the first (except for arsenic) synthetic chemical effective against cancer. Another folic acid derivative, methotrexate, is the first chemical known to have cured cancer patients. Although methotrexate may be useful in therapy for other forms of cancer, the only cancer that has been permanently cured with that drug is choriocarcinoma, an unusual cancer that develops in women after pregnancy.

The *in vitro* mechanism of action of these drugs is relatively simple. They bind to the enzyme responsible for converting folic acid into tetrahydrofolic acid, dihydrofolic reductase. Whether this is the mechanism by which folate antagonists work *in vivo* remains to be established.

For securing appreciable antileukemic activity, the following are structural requirements in a folate analog: (1) an amino group must replace the 4-hydroxyl group of pteroylglutamic acid; (2) the glutamic acid moiety must not be altered because such a modification reduces the antileukemic effect; (3) the therapeutic index of the drug can be increased by substituting a methyl or an ethyl group in position 10 or by halogenation of the benzene ring of the p-aminobenzoic acid moiety.

Most cancer therapists have found that tumors tend to become resistant to the drug administered. The mechanism of this resistance is not always clear, and it is often assumed that the drug is in some way rapidly metabolized. In the case of folic acid antagonists this is unlikely because the two most frequently used antifolic drugs, aminopterin and methotrexate, are not readily metabolized by the body and are almost integrally excreted in feces, urine, and bile. In fact, the mechanism of excretion in the urine is of some special interest. Not only does the drug pass the glomerulus in the glomerular filtrate, but part of the drug is actively secreted in the renal tubules by a mechanism akin to that which promotes secretion of other acids, such as salicylic and p-aminohippuric. Therefore, it is possible to interfere with the excretion of folic acid derivatives by administering such acids.

Then by what mechanism is the sensitivity of the tumor to folic acid reduced? Several mechanisms have been observed *in vitro*. One is increased activity in dihydrofolate reductase. More puzzling is the switch in the affinity of the enzyme for binding of the antagonist. Cancer cells have also been found to become less permeable to the antagonists, and the enzyme is in some cases believed to be sequestered so that it is no longer accessible to the antimetabolites. How much any of these mechanisms participate in the development of the resistance *in vivo* is not clear. The formulas of folic acid and its analogs are presented in Figures 5-4 and 5-5.

Glutamine Analogs The biosynthesis of both purines and pyrimidines requires amidation, and the amide group is donated by glutamine. Therefore, one might expect glutamic acid analogs to interfere with the biosynthesis of both purines and pyrimidines.

Such analogs include O-diazoacetyl-L-serine, or azaserine, and 6-diazo-5-oxy-L-norleucine (DON). These analogs seem to act by inhibiting the enzyme phosphoribosyl pyrophosphate (PRPP) amidotransferase. *In vitro* there is first competitive inhibition between the natural substrate L-glutamine and the analog, but after a certain dose of the analog has been administered, the inhibition becomes irreversible. The mechanism of resistance *in vitro* involves *de novo* synthesis of the amidotransferase. The mechanism of resistance *in vivo* is unknown.

Large doses of azaserine or DON produce necrosis in many tissues—particularly the bone marrow and the intestinal epithelium—but pancreatitis, hepatitis, and nephrosis have also been described.

Purine and Pyrimidine Analogs Between 1950 and 1960 a great deal of research was done on the biochemistry of nucleic acid synthesis. Since purines and pyrimidines are the building blocks of nucleic acids, the research included the systematic dissection of the pathways for the biosynthesis of purine and pyrimidine nucleotides. But behind this important endeavor was also the hope that exploitable metabolic differences between cancer and the normal cell would be discovered. Therefore, several purine and pyrimidine analogs were synthesized.

6-Mercaptopurine (6-MP) is a derivative of hypoxanthine. In hypoxanthine the carbon 6 satisfies its two free valences with an atom of oxygen; in 6-mercaptopurine the oxygen atom is replaced by a sulfur atom. *In vitro* 6-MP is converted to the ribonucleotide, which is known to feedback inhibit the conversion of PRPP to phosphoribosylamine and prevent the conversion of inosinic acid to guanylic acid and adenylic acid. 6-MP therefore interferes with the biosynthesis of all types of RNA and DNA, and actively proliferating tissues are its first target. This may include tumors. Unfortunately, it also includes the bone marrow and the intestinal epithelium.

As for many antimetabolites, the cell protects itself against 6-MP by inhibiting the elaboration of the enzyme which converts 6-MP

FIGURE 5-4. Folic acid.

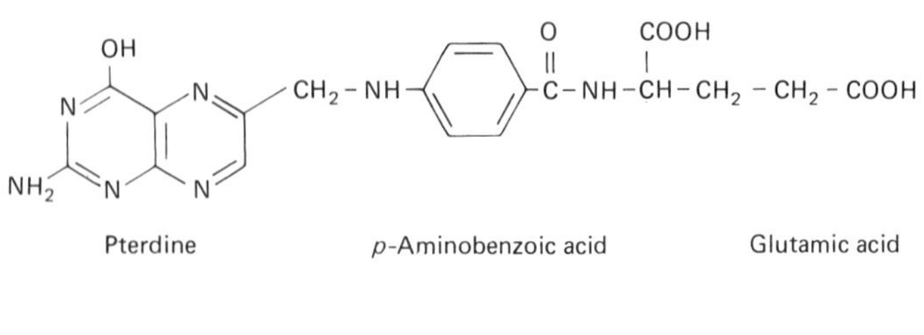

Pterdine   p-Aminobenzoic acid   Glutamic acid

INJURIES TO UNITS OF SPECIFICITY   149

to the ribonucleotide. In an attempt to overcome this resistance, investigators prepared another antimetabolite, 6-MP ribose nucleoside, but it proved to be of little help because it is rapidly converted to 6-MP. Whether 6-MP ribonucleotide will be more effective remains to be seen.

Thioguanine, like 6-MP, has an atom of sulfur replacing the oxygen on the carbon 6. It is an analog of guanine, and like guanine

FIGURE 5-5. Folic acid antagonists.

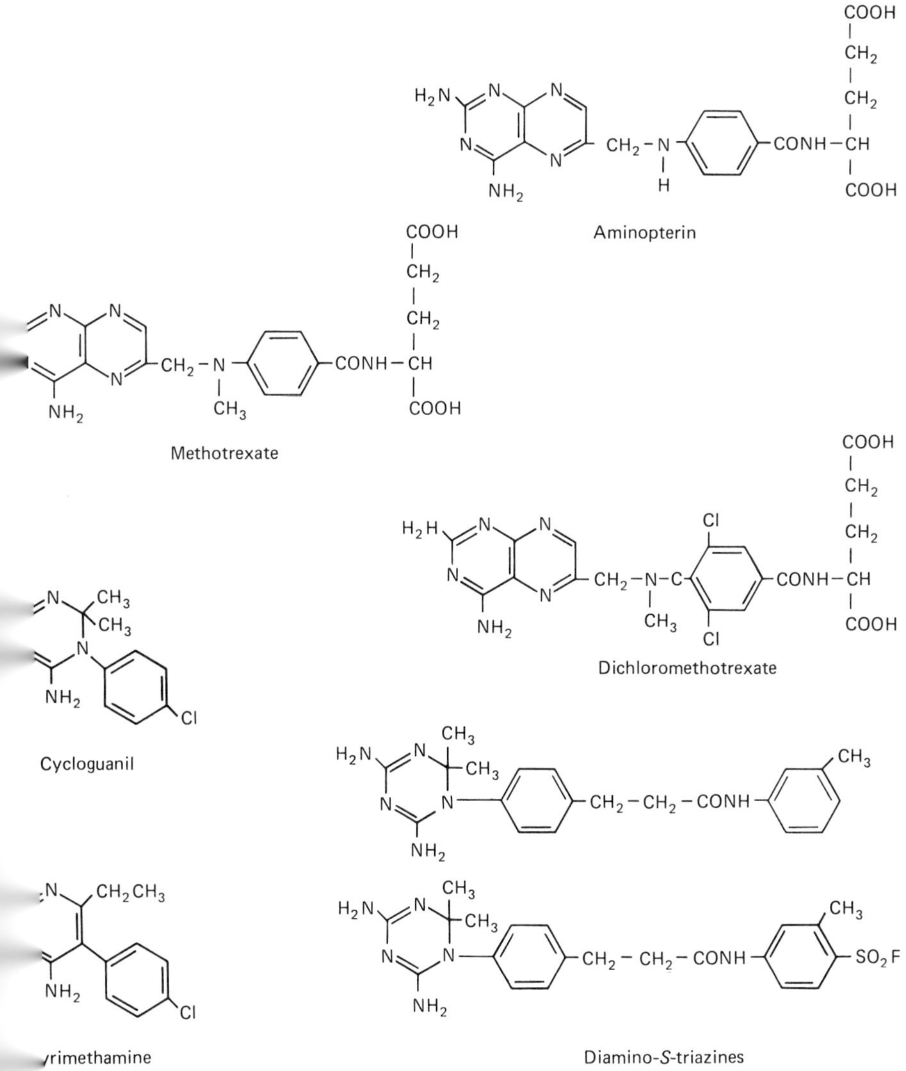

it carries an amino group on carbon 4. *In vitro* its biochemical action is similar to that of 6-MP. Thioguanine is converted to the ribonucleotide and inhibits the same group of enzymes as 6-MP. However, in contrast to those of 6-MP, thioguanine ribonucleotides are incorporated into RNA and are converted to the deoxyribonucleotide, which is further phosphorylated to the triphosphate and incorporated into DNA. Inhibitors of riboside reductase interfere with thioguanine incorporation into DNA. This indicates that thioguanine exerts its toxicity, at least in part, by incorporation into DNA. Thioguanine kills cells in the S phase while DNA synthesis is in process. *In vitro* cellular resistance to thioguanine develops by a mechanism similar to that which developed for 6-MP: namely, deletion of IMP-GMP pyrophosphorylase. *In vivo* the mechanism of resistance is obscure. It is believed to include increased drug metabolism and an inability to convert the ribose nucleotide to the deoxyribose nucleotide.

β-2′-Deoxythioguanosine has properties similar to those of thioguanine except that the former is more effectively incorporated into DNA.

Seldom in the history of therapeutics has a drug been invented with more foresight of its properties than 5-fluorouracil. It was observed that uracil was more readily incorporated into RNA, and, after the proper metabolic conversion, into DNA, of tumors than in that of normal tissues, including some rapidly growing tissues. This was one of the rare exploitable differences between tumor and normal cells. In 1957 Charles Heidelberger, an organic chemist and a biologist who had at that time dedicated his career to the study of cancer, synthesized 5-fluorouracil. The genius of this achievement resided not only in the chemical accomplishment, but also in the logic of his thinking.

Uracil is a direct precursor of RNA, and after conversion to thymidylic acid it is a precursor of DNA. In 5-fluorouracil the hydrogen atom of uracil bound to carbon 5 is replaced by a somewhat similar atom, fluorine. The van der Waals radius of flourine is 1.35 Å, and that of hydrogen 1.20 Å. This should yield an analog of uracil with a stable carbon-fluorine bond. Yet the analog is so similar in structure to the natural substrate that it can be expected to bind to enzymes tha convert it to the ribotide. The ribotide is then converted to the triphosphate, and the riboside of 5-fluorouracil may be further converted to the deoxyriboside. The diphosphoribose fluorouracil and the triphosphate of deoxyuracil riboside should then be incorporated into DNA and RNA (mRNA, tRNA, and ribosomes), and thereby they should interfere with the tumor growth. When tested in *in vitro* systems, all the predictions made proved to be correct.

Normally UMP is converted to dUMP, and dUMP to thymidylic acid by methylation of carbon 5. The presence of fluorine on C5 inhibits the conversion, thereby preventing DNA synthesis. The incorporation into RNA resulted in miscoding of proteins and has interfered with the maturation of ribosomes, but the exact role that the incorporation of 5-fluorouracil into RNA plays in controlling tumor growth is not known.

Once 5-fluorouracil was found to be active against tumors, a number of other derivatives were synthesized for the purpose of testing their activity as antimetabolites; namely, 5-fluoroorotic acid, which is less active; 5-fluorouridine, which is more toxic; and 5-fluorodeoxyuridine, which is less toxic.

Fluoro- derivatives of other pyrimidines were also made: 5-fluorodeoxycytidine and trifluorothymidine (in which the three hydrogen atoms that normally form the methyl group are substituted by three fluorine atoms).

The derivatives have provided agents that act against tumors and also exhibit antifungal and antiviral activity. The latter properties will be mentioned only briefly here.

Although most of the fluoropyrimidine derivatives have antifungal activity, they usually are too toxic to be used as such. 5-Fluorocytosine is an exception; it is relatively nontoxic but retains marked antifungal effectiveness.

On a molar base, F_3TdR is the most effective antiviral agent and can be used topically for herpes simplex and vaccinia virus.

FUdR and FUMP are potent inhibitors of thymidylate synthetase, but the inhibition of these enzymes is not likely to be the only mechanism of interference with tumor growth.

Resistance to fluorinated pyrimidine analogs involves at least three different mechanisms: metabolic breakdown, development of alternative pathways, and cellular adaptation.

Breakdown

FU (but not FC or F_3T) is catabolized to α-fluoro-β-alanine in all tissues, but primarily in liver. Yet the breakdown does not occur in experimental tumors or in human carcinomas of the colon. F_3TdR and FUDR are then degraded to F_3T and FU, which are less effective. FCDR is deaminated to FUDR.

Alternative Pathway

5-FU → 5-FUMP, which inhibits thymidylate synthetase. But in leukemia, the synthetase is bypassed by the direct conversion of thymidine to TMP through the salvage pathway.

Adaptation
Formation of a thymidylate synthetase not inhibited by 5-FUMP. Loss of uridylate kinase, uridine and cytosine kinases, uridine phosphorylase, deoxyuridine phosphorylase, thymidine kinase, and deoxycytidylate deaminase.

In conclusion, the sensitivity of a tumor depends on the balance of its ability to convert 5-FU to FUMP, the active compound, and its inability to degrade FUMP.

5-FU is converted to FUMP in various pathways:

$$5\text{-}FU + PRPP \longrightarrow 5\text{-}FUMP$$
$$5\text{-}FU \longrightarrow 5\text{-}FUDR \longrightarrow 5\text{-}FUMP$$
$$5\text{-}FUMP \longleftarrow 5\text{-}FUDP \longleftarrow 5\text{-}FUTP$$

Cytosine arabinoside (ara C), an analog of fluorocytosine in which the 2′-hydroxyl group is in the *trans* configuration, is especially active in acute leukemia in adults. Ara C is normally converted to the diphosphate and the triphosphate ara CTP. Leukemic cells are rich in kinases capable of effecting such transformations.

The active metabolite of ara C is ara CTP which: (1) inhibits, but only mildly, the reduction of CDP to dCDP; (2) blocks the incorporation of dTTP into DNA through DNA polymerase; (3) is partly incorporated into small molecular RNA (and there is a reasonable correlation between the incorporation of ara C into RNA and cellular death); and (4) is partially incorporated into DNA (but there is no correlation between ara C incorporation into DNA and cellular death). The cells develop resistance against ara C because of its catabolism.

Ara C is deaminated to uridine arabinoside (ara U), which is ineffective. However, the administration of tetrahydrouridine, which inhibits the deaminase, also potentiates the effect of ara C. Loss of deoxycytidine kinase prevents the conversion of ara C to ara CMP, and the loss of deoxycytidylate kinase prevents the conversion of ara CMP to the active compound ara CTP. Ara C is a potent bone marrow depressant most active in adult myelogenous leukemia. The formulas of glutamine, purine, and pyrimidine antagonists are shown in Fig. 5-6.

Alkylating Agents In considering the effect of alkylating agents, one must distinguish between their binding to small molecules and their binding to macromolecules. As we have pointed out, alkylating agents react with any nucleophilic center, including primary, secondary, or tertiary amines; sulfides; and heterocyclic nitrogen rings.

INJURIES TO UNITS OF SPECIFICITY   153

Unless the small molecule is very scarce, it is unlikely that combination with the alkylating agent will cause severe damage. Indeed, the alkylated molecule is rapidly excreted and probably quickly resynthesized. In fact, the binding of alkylating agents to small molecules may be one of the mechanisms of detoxification.

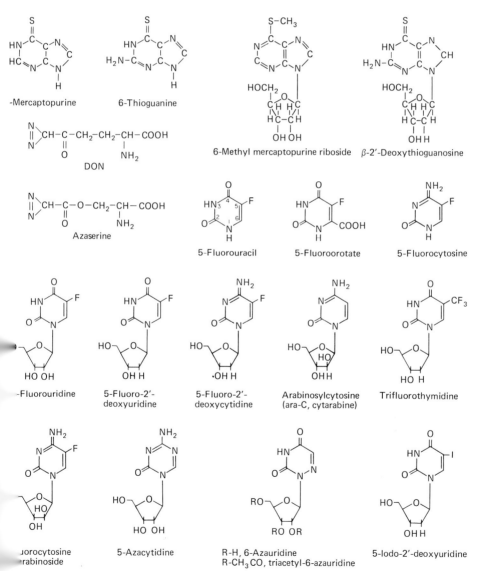

FIGURE 5-6. Glutamine, purines, and pyrimidine antagonists.

Such potential detoxification was taken advantage of during World War II when the British anti-lewisite (BAL) was delivered to every soldier for the purpose of protecting the skin and the mucosa that might be exposed to nitrogen mustard gas.

The binding of the alkylating agent to macromolecules is probably of much more biological significance than its binding to small molecules. Among the macromolecules to which alkylating agents bind are proteins and nucleic acids. Some structural proteins undoubtedly bind alkylating agents, but it is generally believed that nucleic acids are the primary target of alkylation and that within the nucleic acid the purines and the pyrimidines are favorite moieties for their attack.

To simplify the matter, it may be useful to distinguish between monofunctional and polyfunctional alkylating agents. The most frequent site of binding of a monofunctional alkylating agent is the N7 of guanine. Much less alkylation occurs at position 1 of adenine, position 3 of adenine, or position 3 of cytosine.

Polyfunctional alkylating agents generate cross-linking. The cross-linking may be intrastrand and link two N7s of guanine within a strand, or it may be interstrand and link two N7s of guanine in each individual strand, thereby building an interstrand bridge or cross-link. Alkylating agents may also bind to the phosphate group of the nucleic acid. Such binding forms a triester so reactive that it is rapidly hydrolyzed. Some investigators have, however, suggested that the alkyl group bound to the phosphate group is transferred to purine and pyrimidines. We have seen that alkylating agents may bind to small or large molecules and that binding to any molecule could cause some of the biological effects that are manifested by the alkylating agents. These effects include mutations in bacteria, congenital anomalies, and cell death in animals and humans.

The monofunctional alkylating agents such as ethylmethane sulfonate, ethylene amine, and diethyl sulfate are more effective mutagens than the polyfunctional alkylating agents. Moreover, there seems to be a poor correlation between the alkylation of the 7 position of guanine and mutation, whereas the correlation between the alkylation of the 6-O position of guanine of DNA and mutagenicity is good. Monofunctional and polyfunctional alkylating agents are carcinogenic.

The most potent carcinogenic alkylating agents are the nitrosoureas and the triazines, which may cause a high incidence of tumors of the nervous system and the brain. The mechanism of carcinogenicity is not known. All alkylating agents—particularly nitrogen mustard, aziridines, busulfan, and nitrosourea—are teratogenic in animals and humans. Finally, alkylating agents can cause cellular death.

It is generally believed that whereas the monoalkylating agents primarily cause mutations and carcinogenesis, the polyfunctional agents responsible for cross-linking of DNA cause cellular death. The biochemical mechanism by which alkylation exerts its cytotoxic effect is not clear. The most striking biochemical alteration is interference with DNA synthesis, which is usually greater than interference with RNA or protein synthesis or with glycolysis and respiration. It is therefore believed that the alkylation of protein contributes little to the cytotoxic effect, although that of DNA is responsible for the block of DNA synthesis. Alkylation of DNA indeed leads to depurination and breakage of individual strands. It is also possible that damage to RNA interferes with transcription. The formulas of some alkylating agents used in chemotherapy are shown in Figure 5-7.

INJURIES TO TRANSCRIPTION

To provide each cell with its exact protein mosaic, each producing its proper catalytic or structural function, the cell DNA must not only be transcribed faithfully, but the transcribed RNA must ultimately attach to ribosomes and be translated into a polypeptide chain.

Transcription involves copying of the DNA sequence into an RNA sequence by an RNA polymerase. Transcription can thus be inhibited in at least two ways: direct damage to DNA and inhibition of RNA polymerase. Ionizing radiation and actinomycin D cause direct damage to DNA. α-Amanitine inhibits RNA polymerase.

We have already mentioned what kind of damage occurs after ionizing radiation. Actinomycin D is an antimetabolite extracted from a fungus. The molecule is too toxic to be used in cancer therapy except under special conditions. It intercalates between the DNA strands and binds to a guanine residue, so it blocks the RNA polymerase that attempts to transcribe DNA. The cell can recover after receiving small doses of actinomycin D, but it dies if large doses are administered.

Mammalian cells contain three different types of RNA polymerases: two are found in the nuclear sap and one is found in the nucleolus. α-Amanitine, a toxin extracted from a mushroom, inhibits one of the RNA polymerases found in the nuclear sap. When injected in animals, it causes extensive cellular death in the liver.

FIGURE 5-7. Alkylating agents used in chemotherapy.

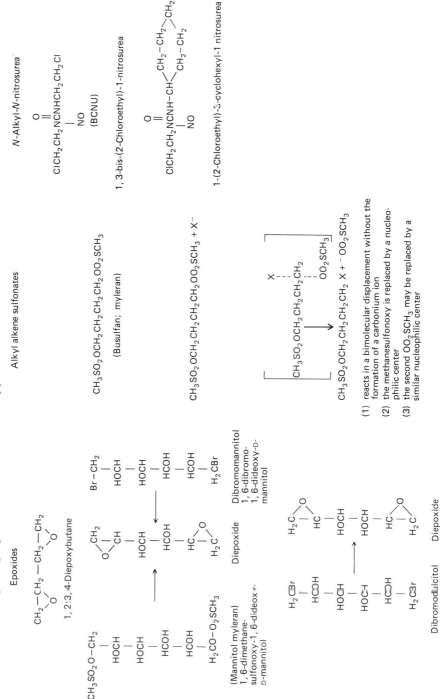

INJURIES TO UNITS OF SPECIFICITY 157

(Mitomycin C)

(Portiromycin)

Activation requires
(1) enzymic reduction of the quinone
(2) elimination of the methoxy group
(3) decomposition of the carbamyl group
Yielding in carbonium ion
There are thus two alkylating centers making it a bifunctional agent

(Triethylene melamine; TEM)

[2, 3, 5-tris (Aziridino)-1, 4-benzoquinone; trenimon]

(N, N', N''- Triethylene thiophosphoramide; thio-TEPA)

(1) in acidic media the nitrogen atoms of the aziridines becomes protenated with resulting increasing alkylating activity
(2) the rings open
(3) and a 2 substituted ethylamine or ethylamide forms

$ClCH_2CH_2-N-CH_2CH_2Cl$

(Nitrogen mustard; HN_2; mechlorethamine)

$ClCH_2CH_2-N-CH_2CH_2Cl \rightleftharpoons ClCH_2CH_2-N^+ \quad + Cl^-$

$ClCH_2CH_2-N-CH_2CH_2OH \quad ClCH_2CH_2-N-CH_2CH_2X$

(1) cyclization with chloride ion generation and positively charged athylenimmonium ion
(2) combination of ethylenimmonium ion
(3) cyclization of the second 2-chloroethyl group to form ethylenimmonium ion

Although excessive breakdown of messenger RNA may not truly result from interference with transcription, the consequence of mRNA breakdown would be the same as a block of transcription if all mRNA were destroyed. Thus, if accelerated breakdown of mRNA occurs as a cause of disease, it is likely to be selective. Some forms of thalassemia, a type of hereditary anemia, might result from accelerated breakdown of messenger RNA. The victim is unable to synthesize one polypeptide chain (α or β) of hemoglobin. At least in some cases of thalassemia, it is believed that this block is due to the elaboration of a messenger RNA which is unusually sensitive to breakdown enzymes.

PATHOLOGY OF TRANSLATION

A protein is elaborated in a multistep sequence. DNA is transcribed into RNA, and RNA is translated into protein. Translation is a complex macromolecular event involving ribosomes, RNA, and several enzymes and cofactors.

In essence, translation involves binding of messenger RNA to ribosomes to form polyribosomes. The polypeptide chain is initiated by recognition of an initiation codon (3 base sequences on the mRNA by the initiation anticodon, formylmethionine tRNA in bacteria and methionine tRNA in mammalian cells). The methionine tRNA on the ribosomes is transferred from a donor site (where the tRNA donates its amino acid to the polypeptide chain) to an acceptor site (where the tRNA loaded with the first amino acid is prepared to form a peptide bond with the second amino acid of the polypeptide chain carried by a second molecule of tRNA). The transfer of tRNA from donor to acceptor site is a complex event involving GTP and numerous protein factors. Normally it goes on step by step until the entire message is translated. When translation is completed, the new polypeptide chain is detached from the ribosomal support, and a new but similar protein can be made on the polyribosome, or the messenger can be degraded.

Numerous antimetabolites secreted by bacteria or fungi—including chloramphenicol, puromycin, and cycloheximide—have been found which interfere with one or more individual steps of translation. The addition of these antimetabolites to bacterial or mammalian cell cultures or their administration *in vivo* interferes with the biosynthesis of all proteins. Depending on the dose, the cell either recovers from the block of protein biosynthesis, possibly by metabolizing or eliminating the antimetabolite, or it is killed. *In vivo* those cells most active in synthesizing protein, such as the hepatocyte or the cells secreting the zymogen granules in the pancreas, are most sensitive to the effect of inhibitors of protein synthesis.

The tRNA molecule is shaped somewhat like a cross; two arms of the cross have a major functional role in protein synthesis. One carries the amino acid, the other carries the anticodon. The anticodon is made of three bases that are complementary to a base triplet on the mRNA. There are only 21 amino acids to be inserted in the protein molecule, but there are many more tRNAs. Therefore, each amino acid has more than one tRNA, and sometimes as many as six.

As already mentioned, the elaboration of a new protein involves the translation of a molecule of mRNA composed of only four different bases in a polypeptide chain that may include as many as 21 amino acids. The codon in the messenger RNA is a base triplet (e.g., ACU). The triplet is recognized by an anticodon located on tRNA (e.g., UGA if the codon is ACU).

Whether or not the wrong amino acid can be inserted on the amino end of a tRNA molecule is not known. But if such a mistake occurred, it would result in the elaboration of a polypeptide chain with an abnormal amino acid sequence. The consequences for the cell's economy would depend on where in the sequence the stowaway amino acid was inserted. If it were in a fraction of the molecule that was not important to the protein's function, then there might be no way to find out that an abnormal polypeptide had been synthesized. On the contrary, if it were in a fraction that involved the active center of an enzyme, the catalytic properties of the enzyme might be modified.

The description of protein synthesis has been oversimplified. In reality, the elaboration of the polypeptide chain involves several steps and numerous factors. Interference with any of these steps blocks protein synthesis. The steps include: (1) initiation, which requires a special tRNA that binds to ribosomes; (2) elongation, which occurs by displacement of the first tRNA to another position of the ribosome making the original place free for a second tRNA to bind; and (3) formation of the peptide bond and termination of the polypeptide chain. Many antibiotics have been found to interfere with one or more steps in protein biosynthesis. Only a few examples will be given. Puromycin competes with acyl-tRNA for its binding site on the ribosome; chloramphenicol inhibits the binding of mRNA to the ribosomes; and cycloheximide inhibits peptide elongation.

The structural support for protein synthesis, the endoplasmic reticulum, is composed of membranes to which ribosomes are attached. The ribosomes are linked together by a ribbon of mRNA to form polyribosomes.

At present too little is known of the composition of the membranes of the endoplasmic reticulum to be able to describe alterations that could interfere with protein synthesis. Nevertheless, it is not unlikely that either hereditary or acquired damage to these structures could interfere with protein synthesis. Tannic acid, for example, causes disaggregation of polysomes, and the endoplasmic reticulum is modified and denuded of its ribosomes. Carbon tetrachloride releases glucose-6-phosphatase from its binding to the membrane of the endoplasmic reticulum. Carbon tetrachloride also interferes with protein synthesis by disrupting the polyribosomes, and ^{32}P uptake increases in the membrane of the endoplasmic reticulum during the first 2 hours after the injection or administration of carbon tetrachloride.

Protein synthesis may also be interrupted by introducing an amino acid analog in the body of an animal. Ethionine is an analog of methionine. Although its mode of action in the liver partly involves its conversion to *S*-adenosylethionine, a compound in which ATP is trapped, in the pancreas ethionine is more likely to interfere with translation. The mechanism of interference is, however, not clear.

The bacteria that cause diphtheria harbor a bacterial virus (called a phage), which elaborates a toxin. The diphtheria toxin is a polypeptide composed of an A and a B fragment. The B fragment enters the cell where it catalyzes the breakdown of nicotinamide adenine dinucleotide (NAD) to yield nicotinamide and ADP ribose. ADP ribose binds to one of the enzymes involved in protein synthesis (transferase 2) and thereby interferes with protein synthesis.

Low doses of puromycin, cycloheximide, and ethionine in mammalian cells result in a temporary block of protein synthesis. This temporary block is often associated with focal cytoplasmic degradation, a process that we will discuss in more detail later.

Ribosomes are made of RNA and proteins. The RNA is syn-

FIGURE 5-8. Types of mutations.

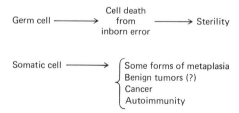

thesized on DNA templates, and the proteins are elaborated by transcription and translation. Although to our knowledge no lesions are attributable to hereditary or acquired alterations in the transcription of the ribosomal RNA or to the transcription and translation of ribosomal proteins, it is not inconceivable that such lesions occur.

Interference with translation resulting from the action of antimetabolites or toxins or from genetically abnormal ribosomes could yield small polypeptide chains with unexpected pharmacological properties.

CONCLUSION

The difference between a liver and a kidney cell resides in its molecular composition. Small molecules are either found in the environment or generated through metabolism. Large molecules are made in the cell. The macromolecules playing the key role in determining cellular specificity are enzymes. Enzymes are proteins synthesized in the endoplasmic reticulum using: messenger RNA as a blueprint, transfer RNA for translation of the code stored in the messenger, and polyribosomes as scaffolds bringing all the components needed for protein synthesis together.

The messenger RNA is transcribed from one strand of DNA which is the repository of all genetic information. Damage by physical, chemical, or biological agents to DNA, if not integrally

FIGURE 5-9. General pathogenic mechanism.

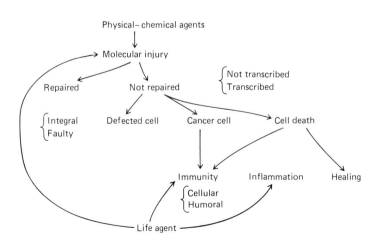

TABLE 5-2. Damage to DNA

Agent	DNA		
A. *Base alterations*			
UV radiation X-Rays Chemicals Carcinogens	1. Mistakes in copying 　　During semiconservative replication 　　During repair 2. Changes in the ratio 　　of keto groups on bases 　　Amino groups on bases	A. Repair B. No Repair 　　1. Nonsense mutation 　　　　No transcription 　　　　Polypeptide deletion 　　2. Missence mutation 　　　　If transcribed abnormal polypeptide	
B. *Crosslinks*			
Carcinogens U.V. Acridine Mitomycin X-Rays	1. Side Chain, e.g., acetylaminofluorene 　　on C_8 of guanine 2. Intrastrand, e.g., thymine dimers 3. Interstrand, with proteins (e.g., 　　thymine and cysteine)	A. Repair 　　By excision repair 　　Excision repair B. No Repair 　　　　Death	Interference with DNA synthesis: Transcription Transformation Mutation	Maldifferentiation

C. *Strand breaks*

   A. Single   A. Repaired ligase   No lesion
       1. PO₄ ester bonds   B. Not repaired

Ionizing radiation
UV radiation   2. Ring rupture (base or sugar)   A. Repaired by   Exonuclease
Virus   Endonuclease   No lesion
Alkylating agents   B. Not repaired   Cell death
                                                                                  Mutation

                                                                      Single   Multiple   Autoimmunity
                                                                                                                                               ?

   B. Double   A. Repairable   No lesion
                                                                       B. Not repairable   Deletion
                                                                                                                Translocation
                                                                                                                Insertion of foreign DNA
                                                                                                                         (Virus)

TABLE 5-3. RNA Synthesis

Cause of damage	Molecular injuries		Consequences of injuries

Indirect	A. Damage to DNA interference with transcription of:	1. Messenger RNA
		   No messenger
		   Abnormal messenger
		2. Ribosomes RNA
		   No ribosomes?
		   Long-lived ribosome
		   Short-lived ribosome
		3. Transfer RNA
		   No transfer
		   Long-lived transfer
		   Short-lived transfer
			Cell death
			Distortion of gene expression
			Short-lived in RNA
			Long-lived in RNA (excess protein?)
			General distortion of protein synthesis
			
			?

	1. Messenger 2. Ribosomal 3. tRNA 4. All three		
Direct			
X-Rays Carcinogens Hormonal imbalances Vitamin deficiencies	Alteration of repression Derepression mechanism	mRNA rRNA tRNA	Distortion of Gene Expression a. E.g., sex characteristics in sex b. Metaplasia in vitamin A deficiency c. Defective maturation in B_{12} deficiency Malignant Transformation

TABLE 5-4. Translation

Cause of damage	Molecular injuries	Consequences of Injuries
See above	Injury to: DNA RNA (m, r, t)	Block in protein synthesis Excess protein synthesis Decreased protein synthesis
Antimetabolites Puromycin Cyclohexamide Chloramphenicol	Interference with assemblage Binding of mRNA to ribosome Binding of tRNA to ribosome Detachment of polypeptide chain	Block of protein synthesis death?
Injury to DNA or transcription	Alteration of ribosomal proteins Not synthesized Abnormal protein	Block of protein synthesis death?
? ?	Enzymes for protein metabolism Anabolic Catabolic	? ?
Antimetabolites (ethionine)	Bioenergetic pathway	Block in protein synthesis
CCL4 Tannic acid	Alteration of membranous support for protein synthesis	Block in protein synthesis
Chronic insults	Distortion of regulatory loops?	Decreased protein synthesis in diabetes

repaired (Table 5-2), interferes with transcription (Table 5-3), or translation (Table 5-4) and causes alterations in protein synthesis which may include: the synthesis of abnormal protein, decrease or excess in protein synthesis, deletion of one or more proteins.

Depending upon whether a germ cell or somatic cell is injured (Fig. 5-8), the ultimate manifestation at the cellular and the organ levels may involve defective function, cancer, or cellular death followed by inflammation healing and sometimes immunological reactions (Fig. 5-9).

INJURIES TO CATALYTIC UNITS

ABOUT ATOMS AND MOLECULES

In previous chapters we have stated that atoms combine to make molecules, molecules interact to make larger molecules (macromolecules), and macromolecules with different properties complex to make organelles, such as chromosomes and mitochondria. We made it appear as simple as building a toy house with wood blocks. Obviously, the interaction of atoms and molecules is much more complicated; so complicated that a description of the details is far beyond the scope of the book. Yet I shall try to give the reader a taste of and possibly an appetite for learning about these complex interactions.

To appreciate such complexity, one must consider the structure of the atom. Atoms are made of a nucleus and peripheral electrons. The nucleus is composed of at least two different types of small particles: protons and neutrons. Both particles have similar masses and sizes, but they differ in their electric charge. Although the neutron carries no charge, the proton is positively charged. If the atom is to be stable, the positive charge must be neutralized. This is achieved by surrounding the positively charged nucleus with a cloud of negative electricity. The charges are condensed in packages, called electrons, which combine the properties of a wave and those of a particle and move around the atoms at terrific speed.

Electrons are 1800 times larger than neutrons, but they have a very low density, and their mass is 1800 times less than that of the neutron.

Because the positive charges of the proton are neutralized by the negative charges of the electrons, the atom has a neutral charge. Thus, in an atom the number of protons is equal to that of electrons.

Consider the hydrogen atom. Its nucleus, in contrast to that of all other elements, has no neutrons and only one proton. The positive charge of the single proton is neutralized by a single electron.

Atoms more complex than hydrogen have more protons, and consequently more electrons are needed to neutralize the positive charges. These electrons gravitate along more or less defined orbitals. The first orbital contains two electrons. The intermediary orbitals contain eight electrons. The most peripheral orbital may be incomplete. The various electron subshells are referred to by the letters s, p, d, and f. The electronic orbitals of the sodium atom are shown in Figure 6-1.

The incompleteness of the peripheral orbital determines what is called the chemical affinity between two atoms, or the electronic composition of the peripheral orbitals determines whether two atoms will combine to form a molecule. Sodium (Na) is an atom, chloride (Cl) is an atom, and sodium chloride is a molecule. Both the sodium and the chloride atoms have three electronic orbitals

FIGURE 6-1. Electronic orbitals of sodium.

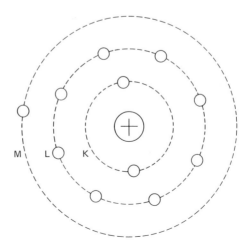

surrounding the nucleus. The first layer contains two electrons, the second layer eight electrons. In the third layer, the most peripheral one, there is only one peripheral electron in the sodium atom and seven in the chloride atom. If the two atoms come together, they will seek to cooperate to balance the negative charges in the peripheral electrons and the positive charge in the nucleus. The chloride accepts the solitary peripheral electron of sodium which avidly donates its electron.

This marriage of convenience between two atoms makes a molecule of salt. The new peripheral orbital common to both the sodium and the chloride atom is called a molecular orbital. It contains eight electrons, or an octet (Fig. 6-2).

Although Bohr assigned rigid trajectories to the traveling electrons, the German physicist Heisenberg postulated that because electrons are light, fast-moving masses, it is impossible to determine their energy and position at a given point in time. This is known as the principle of uncertainty.

However, the studies of Louis Victor de Broglie later established that the electron has, in addition to properties of a particle, properties of a wave. (The wave properties of the electron make it possible to use a beam of electrons in the electron microscope in

FIGURE 6-2. Combination of an atom of sodium and one of chloride to form sodium chloride.

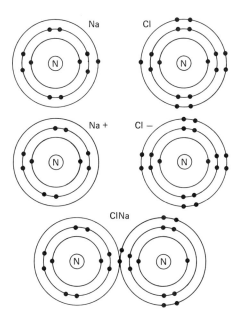

essentially the same way as a beam of light is used in the light microscope.) Schrödinger calculated the wave equation of the electrons. The Schrödinger equation is a differential equation—one that has an infinite number of solutions within defined limits—describing the properties of the electron as a function of its kinetic and potential energy. Consequently, it cannot tell us where the electron is at a given point in time, but it can tell us the boundaries of the area in which an electron with a certain set of energies is likely to be found.

In the hydrogen atom this area is a sphere. But as atoms become more complicated and the electronic shells are more numerous, the shapes of the electronic orbitals vary.

It is difficult to leave such a fascinating subject, but to go deeper into it would require a review of the history of 19th- and 20th-century physics. For further details, refer to Gamow's book, *Thirty Years That Shook Physics*.

Obviously, the chemical properties of atoms and molecules are determined by their peripheral electronic structure. Consider the properties of isotopes. Isotopes of elements have the same atomic number (number of protons), number of electrons, and chemical properties as the original elements. Isotopes differ only by the mass of the nucleus. For example, chlorine with an atomic weight of 355 is a mixture of two isotopes, both containing 17 electrons, 17 protons, but one containing 18 neutrons and the other 20 neutrons.

To form molecules atoms are linked by chemical bonds. There are at least three major types of chemical bonds: ionic, covalent, and metallic. Let's return to the now familiar NaCl. If electrodes are dipped in a solution of sodium chloride and an electric current is passed through the electrode, sodium moves toward the anode (negative electrode) and chlorine moves toward the cathode (positively charged electrode). In the process NaCl is separated into two charged components called ions. Sodium is a positively charged ion (cation) and chlorine is a negatively charged ion (anion). These charged atoms, or ions, are formed during the formation of sodium chloride because the metal has a low ionization potential and gives away its electron readily, while the chloride is avid for electrons. In the molecule the electron of sodium is transferred to the chloride; thus, sodium is positively charged (cation = Na − electron), and chloride is negatively charged (anion = Cl atom + electron). When the marriage is dissolved by electrolysis, the chlorine keeps the electron it gained, and the sodium loses it until another sympathetic chlorine ion meets it and decides to couple with the sodium ion to make a molecule.

In a solution of carbon tetrachloride, a compound often studied by pathologists, if one tries to separate the chloride from the carbon by passing an electric current in the solution, not much happens. The interaction of the valence orbitals (peripheral orbitals) of one carbon and the four chlorine atoms is different from that in sodium chloride. The bonds that link the carbon atom and the chlorine atoms are not ionic but covalent. This is not a marriage of convenience, but a thorough sharing of resources. Linus Pauling, who contributed much to chemistry and was given the Nobel prize in that field* was among the first to describe the nature of chemical bonds.

Carbon tetrachloride is made of four atoms of chlorine which marry one atom of carbon. The carbon has four electron valences, each of which is distributed in one orbital of chlorine. The molecular orbital of carbon tetrachloride is formed by a process in which both atoms give and receive electrons. Thus, the valence shell of each chlorine ends up with eight electrons, one from carbon and seven of its own; and the carbon valence shell also contains eight electrons, seven from chlorine and one from its own. But since no electrons are donated from one atom to the other, no ions are formed. Lewis first described the valence shell in carbon tetrachloride, so this is referred to as the Lewis model.

Methane (CH_4) is made of one atom of carbon and four atoms of hydrogen. The four hydrogen atoms of methane have identical properties, and therefore the bonds that link the carbon to the hydrogen must be of the same length. X-Ray diffraction studies have revealed that the molecule forms a regular tetrahedron with the carbon at the center.

How can such an arrangement be reconciled with the electronic configuration of the carbon atom?

When four atoms of hydrogen combine with one atom of carbon, the molecular orbitals of the carbon atom hybridize. One of the 2s electrons is promoted to a $2p_z$ orbital. As a result, the four orbitals are half-filled:

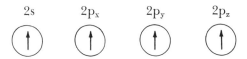

---

* Pauling also founded molecular pathology by his studies on sickle cell anemia. He won the Nobel Peace Prize for his protest against wars because he followed the concern of Lincoln, who said, "Have I no reason to lament for what men do to men?"

The hydrogen atoms share electrons with the four half-filled orbitals to yield four covalent bonds, thus forming a hybrid bond (Fig. 6-3).

Hybrid bonds are stronger than nonhybrid bonds of identical length.

There is another type of bond, the double bond, as is found in ethylene C_2H_4,

$$\begin{array}{c} HC = CH \\ | \quad | \\ H \quad H \end{array}$$

the C-C bond has an energy of 80.5 Cal/mole. Therefore, if both bonds were equal, one would expect that the C=C bond energy would be 161 Cal/mole. In reality, the energy of the C=C bond is 145 Cal/mole. Thus, either both C-C bonds are weakened, or one has normal strength and the other below-normal strength. Available evidence suggests that the latter is true. The weaker bond is called the pi (π) bond.

Double bonds for C=C, C=O, C=N, N=N, and N=O are made of sigma bonds (regular bonds) and a pi bond (Fig. 6-4).

Most metals are hard and have a high melting point, presumably because of a combination of electrostatic attraction and covalent bonding.

FIGURE 6-3. Top: 2p and 2s orbitals of the carbon atom. Bottom left: sp³ hybrid orbitals. Bottom right: tetrahedral orbitals in CH_4.

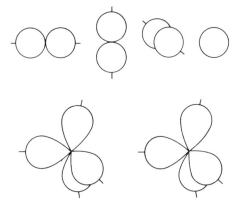

ABOUT CHEMICAL REACTIONS

The rate of reaction is measured by the amount of product generated per unit of time. Guldberg and Waage (1867) discovered the law of mass action, which states that the rate of the reaction is proportional to the product of the concentration of the reacting substances, each expressed to some power, depending on the type of reaction.

$$V = k\ [A] \times [B]$$

Some reactions are reversible, for example:

$$Zn + H_2O \leftrightarrows ZnO + H_2.$$

Reversible reactions eventually reach a state of equilibrium that is always the same whether the reactions start from the left or right. Studies in which the concentration of the reactants was varied revealed that at equilibrium (for a given temperature), the ratio of the product of the concentration of the products to the product of the concentration of the reactive substance is a constant, K.

Thus, in the reversible reaction $A + B \rightleftarrows C + D$,

$$K = \frac{[C]\ [D]}{[A]\ [B]}.$$

In the reaction $Zn + H_2O \rightleftarrows ZnO + H_2$,

$$K = \frac{[ZnO]\ [H_2]}{[Zn]\ [H_2O]}.$$

FIGURE. 6-4. Pi bonding in the molecule of ethylene, $H_2C=CH_2$.

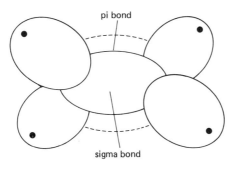

If more than 1 mole of reactant is involved, the equilibrium constant is equal to the product of the concentration of the product and the product of the concentration of the reactant elevated to the power equal to the number of moles involved in the reaction. Thus, in the reversible reaction $2\ SO_2 + O_2 \rightleftarrows 2\ SO_3$,

$$K = \frac{[SO_3]^2}{[SO_2]^2\ [O_2]}$$

The equilibrium of a reversible reaction can be modified in two major ways: by changing the concentration of the reactants or by changing the temperature. But the equilibrium constant is changed only by changing the temperature.

In the reaction $CO + H_2O \rightleftarrows CO_2 + H_2$, an increase in hydrogen displaces the equilibrium to the left. Similarly, if CO were trapped in the course of the reaction, the equilibrium would be displaced to the left.

van't Hoff's equation relates K to T, the absolute temperature in Kelvin degrees:

$$\frac{d\ log\ K}{dT} = \frac{-Q}{RT^2}$$

Q is the heat of the reaction in calories, and R is the gas constant.

The word catalysis technically means breakdown. But the meaning of the word has been changed to indicate "activity because of presence." This notion will become clearer later. Catalysis was first used by Berzelius in 1844 in his report on studies of the formation of ether from ethyl alcohol in presence of sulfuric acid.

$$C_2H_3OH \xrightarrow{H_2SO_4} (C_2H_5)O_2$$

In this reaction sulfuric acid heated at 130°C facilitates the dehydration of ethyl alcohol but does not directly intervene in the reaction, and thus at the end of the reaction as much H_2SO_4 is present as at the beginning.

Catalysts do not change the equilibrium of the reaction; they simply modify its rate. There are two major categories of catalysts: homogeneous, which act in a single phase (as in solution), and heterogeneous, which absorb the reacting molecules at the surface of the catalyst. Although the mode of action of a catalyst is not always known, the absorption of the reactants on the catalyst is believed to bring the reagents in close contact and to modify the structure of their electronic shell in a fashion that reduces the activation barrier.

In many reactions devised by organic chemists, metals (platinum, nickel, etc.) are used as catalysts. The hydrogenation of benzene (C_6H_6) is catalyzed by nickel:

$$C_6H_6 + 3\,H_2 \rightarrow C_6H_{12}.$$

The benezene is (in ways too complex to be described here) absorbed on the nickel, and the product is released from the nickel at the end of the reaction.

In Chapter 1 it was stated that the reaction responsible for the chemistry of life was long believed to be governed by laws different from those that govern laboratory organic chemistry. This view was abandoned when Woehler synthesized urea by chemical methods. Later many histological compounds were synthesized in the laboratory by organic chemists, and it became clear that there was no fundamental difference between organic and biological chemistry.

Although the law of mass action, the equation for equilibrium, applies to biological chemistry as it does to organic chemistry, a number of special requirements are associated with chemical reactions in the human organism.

In cells the concentration of the reactants is often so small that the rate of a given reaction would be extremely slow in the absence of a catalyst. Moreover, the conditions (ionic strength, pH, temperature, etc.) are often inadequate for optimal combination of the reactants. Therefore, biological reactions seldom take place without the aid of catalysts—enzymes. Enzymes must not only accelerate the rate of the reaction, but they must also secure specificity (for example, D-amino acid should never be introduced in a polypeptide chain), and they must be regulated to avoid product shortage or excess. Only the plasticity of the protein molecule can provide a catalyst that fulfills all these conditions.

ABOUT ENZYME REACTIONS

In enzymic reactions the substrate is absorbed on a portion of protein molecule called the active center. The active center is made of a group of amino acids that may be far apart from each other on the stretched-out polypeptide chain, but which are brought together because of the folding of the molecule. Thus, substrate and enzyme form an enzyme-substrate complex, ES, which is dissociated to release the enzyme and the product.

$$E + S \rightleftharpoons ES \longrightarrow E + P$$

The assumption that enzyme and substrate form a complex was made by Michaelis and Menten, and kinetic studies of enzyme activities established it on solid ground.

Michaelis and Menten studied the conversion of substrate under constant pH, temperature, and enzyme concentration. Only the concentration of substrate was gradually increased. When the concentration of substrate is plotted on the abscissa (e.g., in micromoles per milliliter of incubation mixture), and rate of reaction is plotted on the ordinate, the plot yields a curve that rises rapidly at first but tapers off about one-third of the way and reaches a plateau. The curve was explained by assuming that the enzyme and the substrate form a complex.

$$E + S \underset{k_{-1}}{\overset{k_1}{\rightleftarrows}} ES \underset{k_{-2}}{\overset{k_2}{\rightarrow}} E + P$$

Thus, in the enzymic reaction four different velocity constants must be considered: k_1 for the conversion of E + S to ES; k_{-1} for the reverse reaction; k_2 for the dissociation of ES to yield P; and k_{-2} for the reverse reaction. The initial velocity of the reaction $v = k_2[ES]$.

The reverse reaction

$$E + P \underset{k_{-2}}{\rightarrow} ES$$

can be ignored because only the initial velocity is considered—namely, the utilization of the substrate rather than the formation of product. Moreover, the concentration of enzyme ([E]) is always much smaller than the concentration of substrate ([S]).

Briggs and Haldane have assumed that the concentration of the enzyme-substrate complex is not in equilibrium with the substrate or the enzyme concentration, but is constant. Thus, a steady state is rapidly reached, i.e., the rate of formation of the complex equals its rate of breakdown. The concentration of free enzyme equals [E] − [ES]. Inasmuch as the amount of substrate bound to the enzyme is extremely low, the concentration of substrate equals [S]. When the steady state is reached:

$$[S]\,([E] - [ES])\,k_1 = [ES]k_{-1} + [ES]k_2$$

[ES], which is practically impossible to measure, can be extracted from the equation:

$$k_1 [E] [S] - k_1 [ES] [S] = [ES] (k_{-1} + k_2)$$
$$k_1 [E] [S] = (k_{-1} + k_2) [ES] + k_1 [ES] [S]$$
$$k_1 [E] [S] = [ES] (k_{-1} + k_2 + k_1 [S])$$

$$\frac{k_1 [E] [S]}{k_{-1} + k_2 + k_1 [S]} = [ES]$$

$$\frac{[E] [S]}{\frac{k_{-1} + k_2}{k_1} + [S]} = [ES]$$

$$\frac{k_2 [E] [S]}{\frac{k_{-1} + k_2}{k_1} + [S]} = k_2 [ES] = v \qquad \text{(Eq. 1)}$$

The maximum velocity, V, is reached at high concentration of substrate when all the enzyme is converted to the enzyme-substrate complex ES, and $V = k_2 [E]$.

$$\frac{k_{-1} + k_2}{k_1}$$

is a constant, referred to as the Michaelis constant, or K_m. Introducing V and K_m in Eq. 1,

$$\frac{V [S]}{K_m + [S]} = v.$$

If the initial velocity, v, equals half the maximal velocity, V:

$$\frac{V}{2} = \frac{V [S]}{K_m + [S]}$$

$$2 [S] = K_m + [S]$$

$$2 [S] - [S] = K_m$$

$$[S] = K_m \qquad \text{(Eq. 2)}$$

Thus, Eq. 2 tells us that: (1) the K_m value equals the concentration of substrate, which yields half the maximal initial velocity; (2) K_m

can be calculated without knowing the concentration of enzyme. K_m is a constant characteristic for each enzyme for a given concentration at a given temperature, pH, etc.

It is usually too complicated to calculate the K_m and V for the equation.

While

$$\frac{V[S]}{K_m + S} = v$$

is a rectangular hyperbole, inversion of the equation yields a linear plot:

$$\frac{K_m + [S]}{V[S]} = \frac{1}{v}$$

$$\frac{[S]}{V} + \frac{K_m}{V} = \frac{[S]}{v}.$$

If [S] is plotted on the abscissa and [S]/v, on the ordinate, a straight line is obtained which intercepts the ordinate at the value of K_m/V, and the abscissa at the value of $-K_m$. The slope of the line $= 1/V$.

Enzyme inhibitors include those that are nonspecific and act by denaturing proteins and those that act more specifically on one or a few enzymes. Specific inhibitors may act reversibly or irreversibly. When the inhibition is reversible, the inhibitors can be removed from the incubation mixture, for example, by dialysis or chromatography. Diisopropylphosphofluoridate (DFP) is an irreversible inhibitor that binds covalently to the active site of cholinesterase.

Reversible inhibitors are either competitive or noncompetitive. Competitive inhibitors have a structure similar to that of the substrate and compete with the substrate for binding of the active center.

Thus, $E + S = ES$, $E + I = EI$, and the amount of enzyme which binds to the substrate depends on the relative concentration of substrate and inhibitor. When the concentration of substrate is very high, all the enzyme binds to the substrate and the velocity is the same as if no inhibitors were present.

Noncompetitive inhibitors have structures different from those of the substrate. They do not compete for the active site, and the inhibition is independent of substrate concentration.

We have seen that in the absence of inhibitor:

$$v = \frac{V[S]}{K_m + [S]}, \text{ or } \frac{[S]}{v} = \frac{[S]}{V} + \frac{K_m}{V}.$$

In the presence of inhibitor (competitive),

$$v = \frac{V[S]}{K_m(1 + \frac{[I]}{K_i}) + [S]}, \text{ or } \frac{[S]}{v} = \frac{[S]}{V} + \frac{K_m(1 + \frac{[I]}{K_i})}{V}$$

Thus, the value of V is unchanged and the K_m is higher in the presence of inhibitor.

$$K_{m_i} = K_m(1 + \frac{[I]}{K_i})$$

In presence of a noncompetitive inhibitor,

$$v = \frac{V[S]}{(K_m + [S])\frac{1 + [I]}{K_i}}, \text{ or } \frac{[S]}{v} = (\frac{[S]}{V} + \frac{K_m}{V})(1 + \frac{[I]}{K_i}).$$

Thus, the K_m is unchanged but the maximum initial velocity is lowered in the presence of a noncompetitive inhibitor.

TABLE 6-1. Properties of competitive and noncompetitive inhibitors

Property	Competitive	Noncompetitive
Structure	Similar to substrate	Different from substrate
Site of action	Competes with active center	Does not compete with active center
Effect of [S]	Modified by [S]	Not modified by [S]
K_m	Lowered	Unchanged
V	Unchanged	Lowered

Many enzymes are often inhibited by excess substrate. The mechanism of substrate inhibition is not always known, but at least three types have been described: (1) the substrate competes with the active site of another substrate, e.g., water; (2) the substrate modifies the enzyme by removing an activating metal; and (3) two substrate molecules bind to the active site, but neither fits properly on the site, and as a result no product is formed.

Many enzymes require ions (Zn^{2+}, Mg^{2+}) for activity. The presence of these ions increases the rate of the enzyme activity. In addition to metals, many other substances (hormones, metabolites, or even the substrate) may activate the enzyme.

Thus, *in vivo* the regulation of enzyme activity depends on the relative concentrations of inhibitors and activators.

Chemical reactions catalyzed by enzymes are, like all reactions, subject to environmental changes such as pH, temperature, and ionic strength. In the intact body the changes in pH and ionic strength are minimal. But when enzyme activities are measured *in vitro*, optimal activities may vary considerably with pH and temperature.

Some enzymes, such as DNA polymerase, function at optimal pH close to 8; others, e.g., acid phosphatase or β-glucuronidase, function optimally at pH 5. Cathepsins are most active at pH 3. The response of the enzyme to pH is obviously linked to the molecular structure of the protein. It is not clear how a cathepsin with an optimal pH of 3 functions in the cell. At extreme pH some enzymes are destroyed.

The effect of temperature on enzymic reactions is like that on all chemical reactions governed by the van't Hoff equation: the rate of the reaction increases progressively with the temperature. However, if one measures enzyme activity as a function of temperature, there is an optimal temperature for activity because extreme temperatures can destroy the enzyme.

The initial velocity of an enzymic reaction is directly proportional to the enzyme concentration, provided that the molar concentration of the substrate is largely in excess of that of the enzyme.

One can distinguish many groups of enzymes: oxidoreductases, which catalyze the reversible oxidation of such compounds as lactic acid or cytochromes; transferases, which, for example, transfer amino groups from asparate to oxoglutarate to yield glutamate or transfer a phosphate group from ATP to glucose to yield glucose-6-phosphate; hydrolases, which break down phosphate groups, triglycerides, and other compounds; lyases, which split a molecule of water from a compound such as L-malate to yield pyruvate; isomerases, which, for example, convert D-glyceraldehyde-3-phosphate to dihydroxyacetone phosphate; and ligases, which combine coenzyme A to acetate or bind two nucleotides in the polynucleotide chain.

LIFE CYCLE OF ENZYMES

The catalytic unit is composed of three types of molecules: substrate, enzyme, and product. The specificity of the reaction is determined by the amino acid sequence of the enzyme. The specificity of the enzyme may be so restricted as to bind to a single substrate and yield a single product. For example, glucose-6-phosphatase will only bind to glucose-6-phosphate to yield glucose. In many cases the activity of the enzyme may extend to a multitude of substrates and lead to many different products. A case in point is the mixed-function oxidases, which react with a great variety of compounds and hydroxylate them either to detoxify them or to convert them to more active toxic compounds.

The equilibrium of the enzyme reaction is determined by the concentrations of the reactants and not by enzyme concentration, but the rate of the reaction is modulated by enzyme concentration and by a number of other factors including temperature, pH, and allosteric effectors, which affect the enzyme activity. Allosteric effects were discovered only recently. The concept of allostericity proposes that in addition to the active site, the enzyme contains amino acid sequences, allosteric sites capable of binding activators or inhibitors.

Although all enzymes are proteins, many function only if the protein (the apoenzyme) combines to a metal or to a nonprotein organic molecule, which is often a vitamin.

The catalytic unit has its own life cycle which is programmed within the framework of the life cycle of the cell. The three stages in the life of an enzyme are generation, activity, and breakdown.

Enzyme generation is part of the process of differentiation and may be modulated by hormones. Thus, glucose-6-phosphatase activity is very low in fetal liver, but it rises rapidly after birth. Alanine or serine aminotransferase activity is relatively low in hepatic cells but is markedly increased by glucocorticoid hormones. In either case the increase in enzyme activity involves the elaboration of new enzyme molecules. The period of activity is that during which the enzyme reacts with the substrate to yield the product. Sometimes the enzyme must be activated before it can react. This can occur in various ways. Some enzymes—e.g., chymotrypsin, trypsin, pepsin, and many other proteolytic enzymes—are elaborated in a precursor form, and it is necessary to split a portion of the molecule before the enzyme can act on the substrate. Splitting a portion of the enzyme molecule modifies the enzyme conformation and makes the active site available for reaction with the substrate. In other cases the enzyme must be released from inside an organelle to act on the substrate.

The pancreas is composed of two glandular systems: one that excretes its product into the intestine (the exocrine gland), and another that excretes it in the blood (the endocrine gland). The exocrine gland secretes a number of enzymes needed for the digestion of carbohydrates, proteins, and fats in food. These enzymes are not secreted from the cell separately in a water solution. They are packaged in little bags surrounded by a membrane called zymogen granules. When the zymogen granules reach the intestinal lumen, the membrane breaks because it cannot withstand the pH of the environment.

Similarly, polymorphonuclears contain a large number of enzymes destined to digest dead tissues and invading bacteria. These enzymes are also contained in little bags, polymorphonuclear granules. When a polymorphonuclear encounters a bacterium, it engulfs the bacterium within its body, the granules explode, enzymes are released, and if the bacterium is not in some way protected against the enzymes, it is digested.

Some enzymes are attached to membranes such as the endoplasmic reticulum, and the full expression of their activity might require some modification of the relationship between enzyme and membrane. Some enzymes (ATPases) break down ATP to ADP and are strapped to other macromolecules by bonds between two sulfur atoms (S-S bonds). Such ATPases can be activated only by the reduction of the sulfur atoms to -SH groups with the separation of the two sulfur atoms and the release of the enzyme from its macromolecular support. We have already mentioned that enzyme activity can be modulated by coenzymes, metals activators, and inhibitors.

The life span of an enzyme is usually shorter than that of the cell that hosts it because the enzyme molecules are constantly destroyed and resynthesized. One speaks, therefore, of enzyme turnover. The exact turnover of most enzymes is not known, but that of an individual enzyme varies considerably (from 30 minutes to 48 hours). Turnover has been studied primarily for liver enzymes.

PATHOLOGY OF THE CATALYTIC UNIT

On the basis of this panoramic view of the catalytic unit, one can list a number of potential molecular lesions that may occur and interfere with the performance of enzymes: absence of the enzyme molecule as observed in some types of inborn errors of metabolism (e.g., phenylketonuria); increased enzyme synthesis (e.g., porphyria); accelerated enzyme breakdown; alteration of the amino acid sequence in the active site (e.g., some forms of glucose-6-PO_4 dehydrogenase deficiencies); alteration of the amino acid sequence at the allosteric site (e.g., some types of gout); absence of coenzyme (many cases of vitamin deficiencies); inability to bind the coenzyme (special types of inborn errors of metabolism); absence of metals (iron, copper, and zinc deficiencies); inhibition of the active site by specific inhibitors (e.g., cyanide and antimetabolites); and modulation of the allosteric site by compounds present in the environment.

Absence of Enzyme Molecules

Phenylketonuria, a hereditary disease transmitted as an autosomal recessive trait, is characterized by the absence of phenylalanine hydroxylase, the enzyme that converts phenylalanine to tyrosine. Because the enzyme is absent, the substrate phenylalanine accumulates and high levels of phenylalanine are found in the urine and the blood. Tyrosine, the product of the reaction, is never lacking because it is taken in with proteins in the food. Therefore, there is no defect of the synthesis of those proteins that contain tyrosine. Normally, small amounts of phenylalanine are converted into phenylpyruvic, phenyllactic, and phenylacetic acid. In phenylketonuria, this alternative pathway for phenylalanine is much more active than in the normal individual, and phenyllactic, phenylpyruvic, and phenylacetic acid accumulate in the blood and tissues. These metabolites are toxic and they inhibit the activity of some enzymes. Among those inhibited are enzymes involved in the conversion of tyrosine to the pigment melanin. Consequently, patients with phenylketonuria have blue eyes and light skin and hair. Phenylketonurics are mentally retarded, probably because of a toxic compound that interferes with normal myelination.

Defective Enzymes

Glucose-6-phosphate dehydrogenase is an enzyme required for the oxidation of glucose through the hexose monophosphate shunt, is a major source of NADPH. Glucose-6-phosphate dehydrogenase is defective in the red cells of many individuals. The defect may result either from inability to elaborate the enzyme or from synthesis of a defective enzyme. The enzyme is defective because of a mutation that leads to the insertion of the wrong amino acid at the active site. In the absence of glucose-6-phosphate dehydrogenase, the reduction of glutathione through NADH is impaired. The survival of the red cell is dependent on glutathione reduction. The life span of the red cell is reduced in patients with glucose-6-phosphate dehydrogenase deficiency probably because there is not enough NADH to reduce glutathione. Thus, in such patients the red cell breaks down, hemoglobin is released and degraded, bilirubin (a product of hemoglobin degradation) accumulates in the blood, and the patient becomes jaundiced. Patients with glucose-6-phosphate dehydrogenase deficiency have hemolytic anemia.

Increased Enzyme Activity

In porphyria hepatica, large amounts of porphyrin appear in the urine. The concentration of porphyrin may be so great that the urine assumes the color of a rich burgundy wine. The increase in blood and urine levels of porphyrin is associated, for reasons unknown, with abdominal pain and dementia. Porphyria is caused by an increase in aminolevulinic acid dehydrase, an enzyme involved in porphyrin biosynthesis. The increase in enzyme activity results from *de novo* synthesis.

Defective Enzyme Regulation

In gout, uric acid accumulates in the blood and is excreted in excess in urine. The urate salts are soluble at the pH of the blood. When the pH of the medium drops, urates precipitate. The precipitate accumulates in tissues, especially in joints, where it triggers an inflammatory reaction. Although the exact cause of the inflammation is not known, deposition of urates triggers a painful crisis. Again for reasons unknown, the big toe seems to be a favorite site for gout. We do not know why uric acids accumulate; among the possibilities are decreased secretion or decreased breakdown of uric acid and increased production of uric acid.

It is likely that various types of gout exist and that each type is caused by a different metabolic distortion. In one form it is believed that a defective enzyme, phosphoribosyl amino transferase, is elaborated. The enzyme is unable to respond to allosteric inhibitors. As a result, excess purine nucleotides are formed and are broken down to uric acid.

A similar situation obtains in hepatomas. These cancers of the liver are rare in the Western world but common in Asia. Hepatomas of various degrees of malignancy can be produced in mice and rats. Those that are the least malignant are called minimal deviation hepatomas because they resemble the normal liver. In normal liver the biosynthesis of cholesterol from acetate is inhibited by the uptake of exogenous cholesterol. The feedback inhibition is abolished in minimal deviation hepatomas.

Defects in Coenzymes

Some inborn errors of metabolism can be corrected by the administration of large amounts of vitamins. In these diseases a structural defect of the enzyme at the site of binding of the coenzyme is responsible for the inability to bind vitamins properly.

When, however, large amounts of vitamins are administered, enough vitamin is bound to secure sufficient enzyme activity.

Primary cystathioninuria is caused by a defect in the activity of cystathioninase, which cleaves cystathionine to yield cysteine and α-ketobutyrate. In most patients with cystathioninuria, cystathionine excretion decreases after treatment with large doses of vitamin B_6. The B_6-responsive enzyme defect results from a mutation that alters the binding site for the coenzyme.

As already mentioned, vitamins are often coenzymes. Thus, the protein or apoenzyme is inactive unless it is bound to the vitamin or to one of its metabolic derivatives. In humans thiamine (vitamin B_1) is converted to thiamine diphosphate, the active coenzyme. Thiamine diphosphate activates a number of enzymes after binding to the apoprotein, among them pyruvic decarboxylase, α-ketoglutaric decarboxylase, and transketolase. Of these three enzymes activated by thiamine diphosphate, pyruvic decarboxylase plays a critical role in cellular metabolism because it catalyzes the introduction of pyruvic acid into the Krebs' cycle. Thiamine deficiency causes beriberi, a disease not unknown in the West, but most common in the Orient because of the high consumption of polished rice (rice is the main component of the diet in those countries). Removing the husk, which contains most of the vitamin B_1, eliminates the major source of that vitamin. A patient with classic beriberi has edema, an enlarged heart, and polyneuritis.

One can only speculate about the relationship between these symptoms and the vitamin deficiency. Vitamin B_1 is needed for the proper functioning of the Krebs' cycle and consequently for the production of cellular ATP. Therefore, a reduction in the level of ATP could be responsible for the symptoms observed in beriberi. But then why do nicotinamide deficiency and riboflavin deficiency, which interfere with the electron transport chain and ultimately with ATP production, lead to completely different symptoms? An increase of the levels of pyruvic acid in the blood (pyruvicemia) has been described in patients with beriberi. Although alternative mechanisms have been proposed, pyruvicemia is likely to result from interference with the use of pyruvic acid in the Krebs' cycle. Pyruvic acid or one of its metabolites may be toxic for other pathways and cause cellular damage. If such a pathogenic mechanism were to obtain, it would not be very different from what is observed in phenylketonuria, although the enzymic block in the latter results from an inborn error of metabolism. In both diseases a substrate accumulates which is then transformed into inhibitors of other pathways.

Vitamin C is a water-soluble vitamin present in many fruits, especially citrus fruits. Its deficiency causes scurvy. The major symptom of scurvy is weakness of the capillary wall, which results in increased bleeding tendency; there is no coagulation defect. The role of vitamin C in metabolism is only partially known. It probably acts as a reducing agent in the conversion of lysine and proline into hydroxylysine and hydroxyproline.

Hydroxyproline is an amino acid found only in connective tissues, where it participates in cross-linking. In the absence of vitamin C, fibroblasts proliferate normally but for unknown reasons cannot elaborate normal amounts of collagen. Does the inability to convert proline to hydroxyproline automatically inhibit collagen synthesis? Moreover, collagen devoid of hydroxyproline is more rapidly degraded. Because of this inability to elaborate collagen, vitamin C–deficient animals or humans have impaired wound healing and vitamin C–deficient children have distorted bone formation.

In vitamin C–deficient children calcium deposition is not directly impaired, but since the fibroblasts do not produce the proper matrix for calcium impregnation, the calcium precipitates irregularly and never provides a compact bone mass. Therefore, the bones are soft and friable and fracture easily. In addition, as we have already mentioned, the capillary walls are weak; bleeding occurs under the periosteum, blood coagulates, and the clot is invaded by fibroblasts and impregnated by calcium. The relationship between the impairment of proline hydroxylation and the weakness of the capillary walls is unknown.

A great deal has been learned about vitamin D deficiency in the last decades. The major contributions were in the field of vitamin D metabolism and were made in the laboratory of De Luca. Vitamin D is produced in an unusual fashion. Ultraviolet light hits the skin, which contains sterols, and with the help of radiant energy the sterols are converted to vitamin D. Thus a quantum of radiation energy is converted into chemical energy. This special mode of production of vitamin D explains why vitamin D deficiency is common in those countries where sunshine is scarce, or in countries where men and women cover their entire body to shield it from the sun. Vitamin D is lipid soluble, in contrast to thiamine which is a water-soluble vitamin. Therefore, vitamin D is lost in the stools under any condition that leads to malabsorption of lipids in the intestine.

After it is converted to an active compound, vitamin D facilitates calcium absorption by the intestine. The mechanism by which it exerts this effect is unknown. Rickets, caused by vitamin D deficiency, was so common in England during the early industrial era that the disease was referred to as the English Krankheit. Vitamin D has long been known to facilitate absorption of calcium by the intestines; whether it also facilitates calcium incorporation into bones or reduces calcium excretion in the kidney is still debated.

It has been established that vitamin D is not the factor responsible for increased calcium uptake by the intestine, but that two other metabolic steps are involved. One takes place in the liver: conversion of calciferol to a 25-hydroxycalciferol, and the other in the kidney: conversion of 25-hydroxycalciferol to the active compound.

In vitamin D deficiency the fibroblasts and cartilaginous cells continue to elaborate collagen and fundamental substances, but neither becomes impregnated by calcium. As a result, the membranes (fontanelle) that in children hold the bones of the skull are wider than normal. The long bones are soft, and because of pressure they flatten and widen at the joints. A child with vitamin D deficiency has bowed legs, a distended pelvis, and a curved sternum. In a very young child the cartilaginous portions of the ribs proliferate, forming knobs on both sides of the thorax.

The retina is made of a chain of neurons that connect photosensitive structures to the optic nerve. The impulse perceived when light reaches the photosensitive structure is converted into electrical impulses, which are transferred to a special group of neurons in the brain. Two types of photosensitive structures exist: cones and rods. The first perceive black and white; the second, colors. The photosensitive structures contain vitamin A and a protein opsin. Vitamin A is a complex molecule made of long carbon chains containing several double bonds and some methyl side chains, and it can exist in numerous isomeric forms. The terminal carbon has an alcohol group. Under the influence of light, the alcohol is converted to the aldehyde. The all-*trans* isomer is transformed to an 11-*cis* isomer, which binds to opsin to form rhodopsin. In the dark rhodopsin is split to yield the vitamin A aldehyde, which is reduced to the alcohol and the free protein opsin.

Although it is now clear why vitamin A deficiency causes night blindness, we have no notion why it causes transformation of the cells of the columnar epithelium of the stomach, the transitional epithelium of the bladder, or the malpighian epithelium of the vagina into squamous cells (metaplasia).

Vitamin D, after its conversion to the active compound, facilitates the absorption of calcium by the intestine. The mechanism by which it exerts such an effect is unknown. Vitamin K is needed for the biosynthesis of prothrombin, an essential component of blood coagulation, but is not known how vitamin K regulates the biosynthesis of that compound.

Metal Deficiency

Iron is a central component of many functional proteins, including hemoglobin, myoglobin, cytochromes, catalase, and peroxidase. We shall consider here only its association with hemoglobin. Hemoglobin is a large protein composed of two polypeptide chains, the α and the β chains. Each molecule of hemoglobin contains two α and two β chains, and therefore hemoglobin can be considered a tetramer.

A protein is like a necklace made of 21 different kinds of beads of various shapes. The necklace can be folded to yield different tridimensional shapes (conformations). The conformation of a protein is not haphazard, but it is primarily determined by the amino acid sequence and may be modulated by the attachment of prosthetic groups and the surrounding environment. Not only are the amino acid sequences of both the α and β chains of many normal and abnormal hemoglobins known, but a great deal has been learned about the interaction of the two polypeptide chains and their mode of folding to yield the proper conformation of a certain hemoglobin.

In addition to the four polypeptide chains (two α and two β) hemoglobin also contains a polycyclic compound, porphyrin IX. The unit structure of porphyrin is the pyrrole ring composed of four atoms of carbon and one of nitrogen. The carbons are located at the four angles of a rectangle, and the bonds that link the nitrogen to the carbon ring form a triangle. Thus, the pyrrole ring forms a carbon tower topped with a nitrogen steeple, and four pyrrole rings are hooked together by $-CH$ groups. They are in one plane and form a star-shaped structure called porphin. The hydrogen of the carbons at the periphery of the star-shaped structure are substituted in porphyrin IX by either vinyl or propyl residues. Porphyrin IX is inserted in an enclave outlined by the special conformation of the hemoglobin molecule. Each hemoglobin molecule contains one molecule of iron attached to both the porphyrin ring and the protein (globin). The function of hemoglobin is to provide the cell with the oxygen needed for and to eliminate the CO_2 produced by its metabolism. Thus, hemoglobin is essentially an enzyme that binds two substrates, oxygen and CO_2. The binding and the release of these two gases are associated with the release of oxygen. The reduction of hemoglobin changes the conformation of the molecule. The conversion of ferric to ferrous iron is central to the change.

In iron deficiency, hemoglobin is not synthesized, the red cells are small and have low hemoglobin concentration, and the total number of red cells is reduced. The anemia is hypochromic and microcytic but not hemolytic. Thus, the life span of the red cell is not changed, and porphyrin IX is not synthesized in excess of the globin and then broken down. There is no interference with the proliferation of red cells. On the contrary, foci of erythropoiesis appear in tissues which are normally devoid of erythropoietin after birth, such as the liver and the spleen. Consequently, some mechanism must regulate iron uptake, globin and porphyrin synthesis, and proliferation of red cell precursors. The molecules sensitive to or responsible for this regulatory mechanism are not known. As soon as iron is administered, the precursor of the erythrocytes matures, reticulocytes (red cells already devoid of their nucleus but containing a remnant of endoplasmic reticulum and capable of synthesizing hemoglobin) appear in relatively large numbers in the circulating blood, and anemia is corrected within a matter of weeks.

Enzyme Inhibition

Chemical energy of the cell is provided by molecules of ATP that contain an energy-rich phosphate bond. ATP is produced by two major pathways, glycolysis and the Krebs' cycle. The glycolytic pathway is anaerobic; the Krebs' cycle is aerobic. Thus, in glycolysis glucose is converted to pyruvic acid, which is in turn converted to lactic acid. In the Krebs' cycle the pyruvic acid derived from glucose oxidation, some amino acids, and the breakdown product of fatty acids are oxidized; this oxidation is coupled to the reduction of several metabolites, including that of NAD to NADH. The hydrogen of NADH is transported from one molecule to another in a macromolecular chain called the electron transport chain, and is finally coupled with oxygen to yield a molecule of H_2O. The transport of electrons is associated with the release of energy, which serves to convert ADP to ATP. This process of oxidation coupled to phosphorylation is referred to as oxidative phosphorylation. The terminal component in the electron transport chain is cytochrome oxidase. Cyanide binds to the active site of cytochrome oxidase and thereby prevents its oxidation and the uptake of the electron by oxygen to form H_2O. Consequently, the entire process of respiration is blocked, and each cell is condemned to die in a very short time; the victim of cyanide poisoning dies soon after the intake.

Enzyme inhibition does not always have such drastic consequences for the organism and may only partially interfere with cellular metabolism.

One of the sugars found in milk, lactose, is a disaccharide composed of glucose and galactose. In normal individuals galactose is, after phosphorylation to galactose-1-phosphate, converted to glucose in a series of metabolic steps. In galactosemia one of the enzymes needed to convert galactose-1-phosphate to glucose is missing, and as a result galactose-1-phosphate accumulates. Like glucose-1-phosphate, galactose-1-phosphate binds to phosphoglucomutase, the enzyme that converts glucose-1-phosphate to glucose-6-phosphate, an essential step in glycolysis. The binding of glucose-1-phosphate to phosphoglucomutase is readily reversible. In contrast, that of galactose-1-phosphate is much tighter, and the enzyme-substrate complex is difficult to dissociate.

As a result, the phosphoglucomutase reaction is blocked by the presence of galactose-1-phosphate and glycolysis is interrupted. The blockage interferes with the use of glucose, and the patient with galactosemia develops symptoms of hypoglycemia. It is believed by some, but not proven, that the inhibition of phosphoglucomutase is at least partially responsible for the symptoms seen in galactosemia.

DISEASE AND SUBSTRATE ALTERATION

In some cases of injuries to the catalytic unit, the defect resides not with the enzyme but in the availability of substrate. The substrate may be present in inadequate or excessive amounts. Substrate deficiencies can result from inadequate intake of food (e.g., in kwashiorkor), inadequate endogenous production (glucose in glycogen storage disease), or inadequate use (glucose in diabetes).

Substrate Deficiency

Kwashiorkor is observed among children fed a low-calorie diet made principally of carbohydrates and devoid of proteins. This usually occurs when the child is weaned and fed in an area with such a standard diet, for example in Africa, where manioc—a starchy, tuberous plant—is widely consumed. Normally, dietary proteins are digested by stomach and pancreatic enzymes. Digestion means that the large protein molecules are broken down into their unit structure, the amino acids. Some amino acids can be synthesized by the body's cells; others, called essential amino acids, cannot and must be provided by the diet. Without essential amino acids, new proteins cannot be made and a complex symptomatology develops. Although most of the symptoms probably are caused by the defect in protein synthesis, a direct correlation between the deficiency and the symptoms is difficult to establish.

A decrease in the conversion of substrate, whether it is caused by decreased levels of substrate or decreased enzymic activity, results in a decrease in the levels of the product of the enzymic reaction. If the product whose level is reduced is a terminal metabolite, there will be no consequences; but if the product is a substrate for a subsequent enzyme reaction, that enzymic reaction will lack substrate.

In the case of excess substrate (phenylalanine) or product (phenyllactic, phenylacetic, and phenylpyruvic acid) accumulation, an increase in the level of the compound may or may not have consequences. The effect depends on the concentration of the various enzymes that are capable of metabolizing the compound, the solubility of the substrate, and the renal threshold for the substrate.

An increase of water intake, or even of sodium, far beyond our needs is without consequences. Excess intake of glucose, if not continuous, has no deleterious effect. Even small amounts of cyanide can be taken, provided that the levels of intake do not exceed the ability of rhodanase to convert cyanide into thiocyanide. However, sometimes the substrate levels exceed the ability of the normal converting pathways to consume them, and they may then become substrates for alternative pathways. An example is the conversion of phenylalanine to phenylpyruvic and phenyllactic acid in phenylketonuria.

If there is no alternative pathway for utilization of the biochemical and the mechanisms of excretion are nonexistent or exceeded, the compound accumulates. The consequences of such accumulation vary depending on whether the biochemical is soluble or insoluble under physiological conditions.

Soluble compounds are partly excreted in the urine and accumulate in blood. The increased blood levels may be inconsequential, or they may be toxic for one or more tissues. The mechanism by which excessive soluble metabolites exert toxicity is not always known. As we have seen, in some cases at least, toxicity results from enzymic inhibition.

If the substrate of a defective enzyme is a relatively small molecule that is insoluble under certain conditions (uric acid in gout, oxalate in oxaluria), it precipitates.

When the substrate of a defective enzyme is a macromolecule that cannot be eliminated by the cell, it accumulates in large amounts in that cell and may ultimately cause mechanical damage that leads to cell death. Examples are glycogen accumulation in liver and muscle in the various forms of glycogen storage disease; accumulation of sphingomyelin, cerebrosides, or gangliosides in various forms of lipidosis; accumulation of lipids in the arterial walls in atherosclerosis; and accumulation of lipids in liver after carbon tetrachloride intoxication.

A number of hereditary diseases called lipidoses are characterized by the accumulation of complex lipid in cells. They result from the absence of enzymes involved in breakdown of the lipid. Figure 6-5 illustrates the known lipidoses and the suspected enzyme defect in each. Only Gaucher's disease will be discussed as an example.

In Gaucher's disease cerebrosides accumulate in the liver and

FIGURE 6-5. Accumulating compounds and missing enzymes in lipidoses.

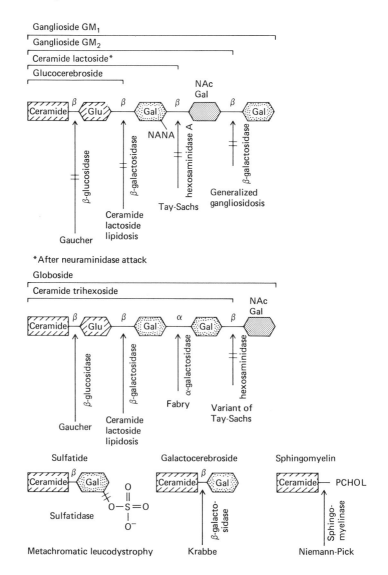

spleen. Cerebrosides are complex lipids composed of a fatty acid, a molecule of sphingosine, and a chain of carbohydrates. One of these cerebrosides, globoside, is a breakdown product of red cell membranes. Normally globosides are broken down to their unit component, the fatty acid, the base, and the carbohydrate. In Gaucher's disease the last breakdown enzyme, a β-galactosidase, is missing. As a result the cerebrosides accumulate in cells that phagocytize red cells, the reticuloendothelial cells of the liver, spleen, and bone marrow. These cells become loaded with the complex lipid.

Atherosclerosis, the major cause of death in developed countries, results from the accumulation of various kinds of lipids in the arterial wall. The lipid accumulation is generally believed to be associated with dietary intake of large amounts of cholesterol. However, it has not been excluded that the intake of other lipids is also significant, and it is also possible that the arterial lesions in atherosclerosis are caused by overproduction of lipids in the arterial walls. Regardless of the pathogenesis of atherosclerosis, lipid accumulation is associated with arterial necrosis and the formation of thrombi causing heart attacks and strokes.

In most cases the exact cause of lipid accumulation in liver cells is not known. Yet the problem is significant because the accumulation of fat ultimately causes the liver cells to die. Cellular death triggers the combination of cellular reactions that lead to cirrhosis of the liver. Three reactions to the injury take place: inflammation, fibrosis, and regeneration. Inflammation is of the chronic type, and its marks are barely visible. Fibrosis varies in its intensity and location in the liver depending on the type of injury. Regeneration is aberrant and the normal histology of the liver is not reproduced. As a result of these reactions, the histological architecture of the liver is totally distorted. The distortion may be so marked that the liver appears abnormal at autopsy and under the microscope. At autopsy the liver is hard and hob-nailed; under the microscope the liver cells are separated by bands of fibrous tissue and the normal vasculature of the liver is distorted.

In glycogen storage diseases, glycogen accumulates in tissue. There are various types of glycogen diseases, and they differ by the nature of the defective enzyme. One type of glycogen storage disease, von Gierke's disease, is caused by the absence of glucose-6-phosphatase. Affected children have large livers and the liver cells are loaded with glycogen, a storage form of glucose. The breakdown of glycogen by phosphorylase yields glucose-1-phosphate, which is then converted to glucose-6-phosphate; this in turn is dephosphorylated before it is made available to muscles as a source of chemical energy. In the absence of glucose-6-phosphatase, glucose cannot be released, and as a result children with von Gierke's disease are prone to developing hypoglycemia. Epinephrine stimulates phosphorylase, the enzyme responsible for glycogen breakdown. In a normal individual the injection of epinephrine activates the breakdown of glycogen, and glucose is released. This does not occur in von Gierke's disease.

Substrate Excess

Excess of a substrate may be exogenous, as is the case in some forms of oxaluria with excess intake of oxalate and possibly in arteriosclerosis with excessive cholesterol intake, or endogenous, as is the case for phenylalanine in phenylketonuria. The excess phenylalanine is metabolized to phenyllactic or phenylacetic acid in an alternative pathway.

CONCLUSION

A number of instances have been described in which the initial injury undoubtedly affects one or more catalytic units because of the absence of enzyme, the absence of coenzymes, a defect in the enzyme in response to effectors or its ability to bind coenzymes. In each case it is usually possible to link the appearance of one symptom—not necessarily the prominent one in the disease—to the enzyme defect. However, in most cases the chain of events that link the initial molecular alteration to the complex set of symptoms, or the link between injury and misery, remains to be discovered. For example, we do not know why the absence of phenylalanine hydroxylase impairs myelination, why the absence of ceruloplasmin leads to copper accumulation with liver cirrhosis and well-localized brain damage, or why galactosemics are slightly icteric and develop cataracts.

Similarly, deficiency of thiamine and vitamins C, A, and D causes a number of symptoms that remain unexplained, such as demyelination, metaplasia of columnar or transitional cells into squamous cells, and breakdown of the capillary walls. An understanding of the pathogenesis of these symptoms is certain to yield new clues to the pathogenesis of still mysterious diseases, such as many of the nervous diseases associated with demyelination or some skin diseases (e.g., psoriasis, which is likely to be associated with defects in maturation of the epithelial cells of the skin). Similarly, the mutation of products and substrates may cause damage, but the molecular target of the injury is unknown.

HORMONE IMBALANCES

ENDOCRINE GLANDS

Many cells of mammals, including man, depend on hormones for the expression of their full activities. Hormones are chemical messengers elaborated by a specific cell found usually in one organ, but sometimes in more than one. Hormones vary in their chemical structure, site and regulation of secretion, and mode of action.

Hormones can be classified in two major groups: nonpolypeptide and polypeptide hormones. The nonpolypeptides include steroid hormones (sex hormones—androgens in males, estrogens and progestins in females—corticosteroids, aldosterone), thyroid hormones, and catecholamines (epinephrine). The polypeptide hormones may be small peptides such as gastrin (a decapeptide) or large polypeptides, e.g., insulin and parathormone.

The adrenal gland, a triangular structure located on top of the kidney, secretes corticosteroid hormones. The adrenal is composed of two parts: the cortex and the medulla. The medulla secretes cortisone, sex hormones, and aldosterone. The cortex is made of three layers of cells, and each layer secretes a different type of steroid hormone. Steroid hormones are also secreted in the sex glands (ovaries and testes). Thyroid hormones are secreted in the thyroid gland, located in the neck at the anterior aspect of the trachea. Catecholamines are secreted by the cells of the adrenal medulla.

The hypophysis is a small, bean-shaped structure contained in a depression of the base of the skull called the sella turcica. The three lobes of the hypophysis are the anterior, posterior, and pars intermedia. The anterior lobe secretes a number of polypeptide hormones that regulate growth and the secretion of thyroxine, sex hormones, and corticosteroids. Each hormone of the anterior hypophysis is secreted by a different type of cell, each of which can be identified with special histochemical techniques.

The posterior lobe secretes a small polypeptide hormone, vasopressin, which regulates water secretion in the kidney.

The pancreas is somewhat unique because it is both an exocrine and an endocrine gland. The exocrine gland secretes digestive enzymes that are packed in zymogen granules and poured out in the intestinal lumen. The endocrine gland secretes insulin and glucagon and pours them directly into blood. Insulin facilitates the penetration of glucose in the muscle cell, and glucagon stimulates gluconeogenesis, the formation of glucose from glycogen, and, when necessary, from amino acids derive from protein degradation. Each pancreatic hormone is secreted by a different cell type: the β cells secrete insulin and the α cells secrete glucagon.

Parathormone is a polypeptide hormone secreted by four small, bean-shaped glands. Parathormone facilitates the release of calcium from bones. Thyrocalcitonin, another polypeptide hormone, is secreted by some specialized cells of the thyroid, and it is still not known how it regulates calcium absorption in bone.

MODE OF ACTION OF HORMONES

During the last three decades a number of major contributions have been made on the mode of action of hormones. They include the elucidation of the amino acid sequence of most polypeptide hormones; the discovery that many hormones are often synthesized in the form of a precursor, which is an extended polypeptide; and the discovery of binding proteins, receptors, second messengers; and the mechanism of hormonal release through releasing factors.

In the 1950s Sanger published the first complete amino acid sequence of a complex polypeptide hormone—insulin. Insulin is made of two polypeptide chains (the A and B chains) held together by S-S bonds. The hormone is synthesized in the β cell of the pancreas in the form of proinsulin. Proinsulin is a polypeptide much larger than insulin in which the A and B chains are joined by a C chain or connecting peptide (Fig. 7-1). Proteolytic enzymes that have not been identified split the A and B chains from the connecting peptide. Another enzyme then connects the A and B chains. A number of other hormones, including glucagon, are now known to be first synthesized in the form of an extended polypeptide chain.

All hormones must be transferred from the site of secretion to their site of action. Polypeptide hormones may be transported free, but many if not all nonpolypeptide hormones are transported in a bound form. Binding usually takes place with blood proteins that are identified by electrophoretic mobilities. Binding of hormones has been most extensively investigated with the thyroid hormones, which are made of iodinated tyrosine. One, two, three, or four atoms of iodine might be bound. Iodination takes place when the amino acid tyrosine is part of a longer polypeptide chain, thyroglobulin. During secretion iodine must be pumped inside the cell

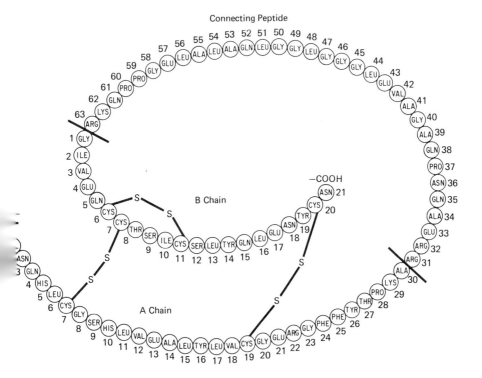

FIGURE 7-1. Proposed covalent structure of porcine proinsulin.

where it binds to the tyrosine of the thyroglobulin. Thyroglobulin is excreted and the polypeptide is partially digested by special enzymes. The iodinated tyrosine is released in the bloodstream. The released hormone binds to either albumin (TPA),* or a binding globulin (TGB).† The binding affinity of the globulin is much greater than that of the albumin, but there is much more circulating binding albumin than binding globulin.

Hormones act at selected sites. For example, insulin stimulates glucose uptake by muscle and adipose tissue but has no effect on brain or uterus; sex hormones stimulate the maturation of gonads and secondary sex organs but have no direct effect on muscle or bone. Such specificity in the site of action of hormones requires that the hormone can recognize the cell on which it acts. Cells susceptible to the action of a given hormone have been found to contain specific receptors. The receptors are likely to be proteins or glycoproteins. They may be located at the surface of the cell, as is the case for insulin receptors, or they may be inside the cell, as is the case for steroid receptors. Thus, insulin bound to sepharose beads, which make it impossible for the insulin to enter liver or muscle cells, can still exert its specific hormonal effects on those cells. This indicates that the insulin receptors must be at the surface of the cell. Estrogens or corticosteroids penetrate the cell membrane, attach to specific protein receptors in the cytoplasm, and are transferred to nuclear receptors before they exert their effect on the target cells.

Cytoplasmic receptors for androgen, estrogen, corticosteroids, and progesterone have been characterized. The cytoplasmic hormonal complex is transferred to the nucleus where it binds to chromosomes believed to contain 5000 "acceptor sites" for hormone-receptor complexes. Although the binding of hormone-receptor complexes with denuded DNA is unspecific, the binding to chromatin seems to be specific. Thus, only the target cells accept the receptor-hormone complex, and the receptor-hormone complex binds to a group of nonhistone chromosomal proteins called AP_3. The mode of transfer from cytoplasm to nucleus is not known.

REGULATION OF HORMONE SECRETION

There are various levels of complexity of the regulation of hormone secretion. In the simplest case hormonal secretion is regulated by the blood levels of one of the metabolites whose usage is regulated

---

* TPA = thyroid hormone bound to plasma albumin.
† TGB = thyroid hormone bound to globulin.

by the hormone; for example, insulin secretion seems to be regulated by the level of glucose in blood. An increase in glucose concentration stimulates insulin secretion, whereas a decrease in glucose secretion impairs it. The mechanism of regulation of insulin secretion by glucose is unknown. Similar to the regulation of insulin by glucose, the blood levels of calcium regulate the secretion of parathormone.

In other cases the secretion of a hormone, e.g., thyroid hormone or corticosteroids, is regulated by a chain of reactions involving several organs. A nucleus in the hypothalamus (an anatomically identifiable group of cells) secretes a releasing hormone, which usually is a small polypeptide. The releasing hormone is carried through the circulation or via the nerves to the hypophysis where it activates the release of a specific tropin (e.g., thyrotropin or ACTH). The tropin is carried in the bloodstream to the endocrine target organ where it recognizes specific receptors: e.g., for thyrotropin in the thyroid, ACTH in the adrenals. The target organ is stimulated by the secretion of the effective hormone, thyroid or adrenal hormones. Those hormones are in turn released in the blood and act on specific nonendocrine target organs (Fig. 7-2).

The secretion of tropins is cut off by the increased levels of

FIGURE 7-2. Schematic representation of hormonal action.

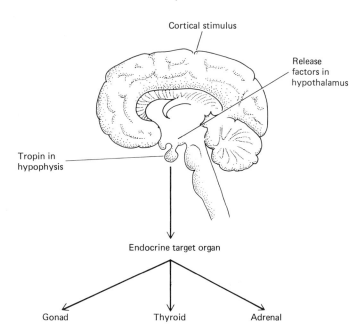

hormone, e.g., thyroxine, estrogens, cortisone (feedback inhibition). Thus, hormone regulation sometimes involves a sequence of steps and several organs: the hypothalamus, the hypophysis, and the target endocrine gland.

SECOND MESSENGERS

How do hormones act on the target organ? We have seen that they become attached to receptors. For the expression of their effect, most hormones require at least two further steps: (1) activation of a second messenger, and (2) activation of a catalytic unit or modulation of gene expression. The second messenger cyclic AMP was discovered in Sutherland's laboratory during his studies on glycogen metabolism. The hormone glucagon binds to a surface receptor of the liver cell. The receptor is located in the cell membrane close to an enzyme, adenyl cyclase, which converts AMP to cyclic AMP. Cyclic AMP activates a protein kinase, which phosphorylates the enzyme that breaks down glycogen, namely, phosphorylase. Consequently, an inactive phosphorylase is converted to an active phosphorylase. Since adenylcyclase can convert more than one molecule of AMP to cyclic AMP, the effect of glucagon on the target cell is markedly amplified. The breakdown of glycogen is interrupted by a phosphatase that dephosphorylates the active phosphorylase.

The activity of α-amino ketoglutaric transaminase is increased in liver cells after the administration of a corticosteroid. In some unknown way, *de novo* biosynthesis of the enzyme is stimulated. The bound hormone is carried by the blood to the target organ where it is released to attach to specific receptors. The steroid hormone, like most other steroid hormones, has a molecular weight of approximately 300 daltons. It can therefore penetrate the cell membrane and bind to a cytoplasmic receptor. It is then transferred from the cytoplasmic to a nuclear receptor probably associated with one of the macromolecular components of chromatin. Binding of corticosteroid hormones to lymphoma cells causes cell death. Receptors for the corticosteroids have been found in lymphoma cells, and their properties resemble those of the receptors found in hepatoma cells in which the corticosteroids induce tyrosine aminotransferase synthesis. It seems, therefore, that in both hepatoma and lymphoma cells the primary reaction between glucocorticosteroid and cell membrane is the same. But thereafter the molecular events triggered by the hormone follow different paths. In one case the hormone kills the cell; in the other it modulates gene expression.

HORMONE DEGRADATION

Hormones, like enzymes, have a life cycle independent of that of the cells that secrete them, or of that of the target cells. After hormones have been synthesized, transported, and have acted on the target organ, they must be degraded; otherwise, they would accumulate.

We know practically nothing of the mechanism of breakdown of polypeptide hormones. Steroid hormones can be inactivated by several metabolic pathways including hydroxylation through the mixed-function oxidases and conjugation of the steroid to a molecule of glucuronic acid to yield steroid glucuronides. The thyroid hormones are deiodinated. Deiodinases have been described, but their regulation is not known.

With this panoramic description of what is known about the mode of action of hormones, it is possible to anticipate what type of cellular or molecular lesions might cause hormonal imbalances.

HORMONAL DISEASES

Absence of Endocrine Organs

Obviously, an essential requirement for hormonal production is that the organ secreting the hormone be present. There are at least three circumstances in which an endocrine gland may be absent: congenital agenesis, destruction by disease, and iatrogenic elimination.

We have already seen how agenesis of the third pharyngeal pouch results in the absence of both the thymus and the parathyroid glands, causing hypoparathyroidism. The parathyroid regulates the level of calcium in the blood. When the blood calcium level is low, parathormone stimulates the release of calcium from the bones. In the absence of parathormone, blood calcium levels stay low.

Some endocrine deficiencies result from acquired injury to the secreting organs. Examples of such injuries are rare cases of diabetes, Addison's disease, Simmonds' cachexia, and Hashimoto's disease.

Diabetes Most cases of diabetes are not caused by obvious damage to the organ. The role of the pancreas in diabetes was discovered by Thomas Willis, who was the first to observe the association between diabetes and the presence of pancreatic stones. Von Meering and Bilchowski demonstrated that pancreatectomy caused diabetes. Best and Banting extracted insulin from the gland and demonstrated that the β cells of the islets of Langerhans secrete insulin. The role of the pancreas in physiology remained unknown until the 18th century.

It is now clear that the pancreas is a complex gland that has both exocrine and endocrine structure. The axis of the exocrine pancreas is the pancreatic duct. The exocrine glands are round or oval-shaped hollow structures lined by an epithelium. The glands are arranged like a bundle of grapes; they open in small ducts that converge to make larger ducts and finally open in the central duct. The glandular structure is lined by epithelial cells that secrete enzymes contained in granules. The enzymes, excreted in the ducts and finally in the intestine, help to digest food. In the midst of these exocrine glands are the islets of Langerhans.

In the islets the cells are not arranged in an acinar form, they do not open in ducts, but they secrete hormones that are directly released in the many capillaries in the islets. As already mentioned, at least two hormones are secreted by the islets: insulin and glucagon.

Diabetes occurs in chronic pancreatitis, some cancers of the pancreas and after pancreatectomy. The pathognomonic symptom in diabetes is the excretion of sugar in the urine. Normally the sugar that passes into the glomerular filtrate is reabsorbed in the kidney tubules. Yet reabsorption by the kidney tubules is limited. Above a certain level of sugar in the glomerular filtrate, the sugar is spilled in the urine. The high sugar content of glomerular filtrate is caused by the high concentration of sugar in the blood (hyperglycemia). Hyperglycemia results from the inability of the peripheral tissue to use glucose. Glucose injected or produced metabolically through gluconeogenesis cannot pass the membrane of the muscle or adipose cell that utilizes most of it, so it accumulates in the blood. Glucose is not used because of the lack of insulin, which in this case results from damage to the pancreas. As will become obvious later, the lack of insulin is not always caused by direct damage to the pancreas.

Hyperglycemia is not likely to be the direct cause of the many and complex symptoms seen in diabetes. Distortions in intermediary metabolism, resulting from the cells' deprivation of glucose, are more likely to cause the symptoms. In addition to hyperglycemia, glucosuria, polyuria, and loss of electrolytes, diabetes is also associated with the development of cataracts, atherosclerosis, and in some cases metabolic distortions too complex to be discussed here. Such metabolic distortions lead to ketosis, or the accumulation of ketone bodies in the blood and tissues. Ketone bodies include hydroxyglutaric acid, acetoacetic acid, and isobutyric acid.

In most cases of diabetes, the lack of insulin does not result from massive trauma to the pancreas as obtains in pancreatitis or after pancreatectomy. Diabetes usually is caused by much more subtle mechanisms of interference with the availability of insulin. Probably not all the mechanisms of interference with insulin activity are known. Known mechanisms include destruction of the β cells of the islets of Langerhans (as occurs in experimental diabetes after alloxan administration or in human diabetes through an autoimmune process), formation of antibodies against insulin, and appearance in the blood of antagonists to insulin. One can only speculate about other potential mechanisms that could cause a lack of insulin, for example, absence of synthesis of the prohormone, synthesis of a defective prohormone that cannot be converted to insulin, absence of synthesis of specific enzymes responsible for conversion of the prohormone to the active hormone, elaboration of abnormal A or B chains, and inability to form S–S bonds between the two chains of insulin. Time will tell if some or all of these potential mechanisms of insulin deficiency exist. Detailed knowledge of the pathogenesis of diabetes would not only provide a clearer disease classification, but it might help also in diagnosis and therapy.

Simmonds' Cachexia The hypophysis may be massively destroyed by thrombosis or invasion by cancer cells. The result is interference with the excretion of all tropins (ACTH, thyrotropin, somatotropin, and gonadotropin) leading to Simmonds' cachexia. The victim presents signs of deficiency in adrenal, thyroid, and sex gland secretion, but the predominant event is a defect in tropin, which leads to atrophy of muscle, skeleton, and viscera. The lack of gonadotropin leads to atrophy of the sex organs. The lack of thyrotropin results in myxedema. In absence of ACTH there is corticosteroid deficiency, and the absence of growth hormone is probably responsible for the severe atrophy of muscle and massive weight loss.

Addison's Disease The adrenals can be destroyed by tuberculosis, causing aldosterone deficiency; Addison's disease results from such a deficiency. Aldosterone is responsible for the reabsorption of sodium in the renal tubules, so Addison's disease is associated with loss of salt in the urine and low salt levels in the blood. Other symptoms include hypotension, increased skin pigmentation, which is unexplained, and severe weakness.

Autoimmune Defects Thyroid deficiency, and probably other hormonal deficiencies, can be caused by self-destruction of the gland in a so-called autoimmune process. In such cases some of the molecular constituents of the gland elicit an immune response that destroys it. Why the immune system is unable to recognize the "self" or its own tissues is not clear. Yet in such an autoimmune defect the glandular tissue is entirely destroyed and replaced by lymphocytes and fibrous tissue.

Nutritional Deficiencies and Goiters

Interference with the secretion of thyroid hormones may result from deficiency of iodine, a compound needed for the synthesis of thyroid hormones. Before iodine was added to comestible salts, this form of thyroid deficiency was frequent in such land-locked countries as Switzerland and some parts of Central Africa. The thyroid gland attempts to compensate for the iodine deficiency by proliferating and secreting large amounts of noniodinated hormones and other substances, and as a result it grows into a large mass or goiter.

Defects in Hormone Synthesis

There is no evidence that hormones are synthesized with abnormal amino acid sequences that interfere with their activity. Perhaps such mutations are lethal, but it would seem unlikely that they never occur. When hormone synthesis requires the formation of a prohormone such as insulin, it is possible that hormonal deficiency could result from a defect in conversion of the prohormone to the effective hormone, either because of an alteration in the polypeptide structure which prevents hormonal breakdown, or because specific enzymes involved in converting the prohormone into the active hormone are missing. Such missing enzymes have not been reported for polypeptide hormones.

Enzyme deficiencies that cause hormonal defects have been described in thyroid hormones causing cretinism and in steroid hormones causing the adrenogenital syndrome. Cretinism is caused by hereditary deficiency of thyroid hormones. At least five different enzyme defects can cause cretinism.

Defects in Hormone Transport

Hormones are transported in the blood by either Aglobulin or albumins. Low levels of binding protein have been described and are believed to be the cause of some forms of hyperthyroidism.

Receptor Defects

We have seen that hormonal action requires that the hormone binds with specific receptors at the surface or inside the cell. Therefore, one may expect that some forms of hormonal imbalance are caused by distortion of the receptors. The receptor could be missing or abnormal. Abnormal receptors could bind either excessive or low levels of hormones. It is possible that the relationship of surface receptors with the membrane could be so distorted that receptors are not accessible to the hormone. We know so little of membrane structure and receptors that one can only speculate on the existence of receptor diseases, but it is likely that several types of such lesions will be described. At least one form of pseudohyperparathyroidism is believed to be caused by a receptor defect.

In breast cancer there seems to be a correlation between the presence of estrogen receptors in the cytoplasm and nucleus and the response of the cancer to estrogen.

Regulatory Defects

Clearly some hormonal imbalances are likely to result from disruption of the delicate mechanism regulating hormonal secretion. Of course, some links in the regulatory chain are themselves endocrine glands (e.g., some nuclei of the hypothalamus or the hypophysis), and therefore they may be susceptible to the kind of damage already described for other endocrine glands. Another possible defect of hormonal metabolism is an inadequate response to the feedback signals emitted by the ultimate endocrine gland or by the blood concentration of substrates. Secondary hypoparathyroidism is an example.

Blood passes through the kidney, which contains a unique microscopic structure, the glomerulus—mainly a bundle of entangled capillaries and a tubule. The tubule is shaped in a hairpin bend, forming an ascending and a descending branch. The straight portions of the tubules are attached to the glomerulus and to large connecting tubules by convoluted tubules. Glomerulus, collecting tubules, and the ascending and the descending loops form the nephrons. In the formation of urine, blood first travels to the glomeruli. The capillaries of the glomeruli are permeable to plasma, but not to cells contained in the blood or to large proteins. The glomeruli filter all soluble components in blood, including electrolytes, amino acids, and uric acid.

Some of the compounds may be excreted directly in the urine, but most are reabsorbed along the tubules. Resorption regulates the quantities of compounds excreted. For example, in the normal kidney most of the calcium filtered in the glomeruli is partly reabsorbed and partly excreted. But in kidney disease, calcium is not reabsorbed and is lost in the urine. As a result the blood levels of calcium decrease. The homeostasis of calcium is rigidly regulated by parathormone. To maintain the calcium levels in the blood, the parathyroid secretes more hormone, causing the resorption of bone calcium. The stimulus to parathyroid causes cellular hyperplasia, and this is known as secondary parathyroid hyperplasia. Although calcium is the major component of bone, calcium ions are also critical in many metabolic steps including membrane interaction and muscle contraction. Unfortunately, we know very little of the molecular mechanisms of calcium action. When the calcium levels of the blood are low the patient develops tetany.

An interesting example of distortion of feedback regulation takes place when an ovary is transplanted in the spleen of a castrated rat. The steroid hormones secreted by the ovary are then degraded by the liver cells. Therefore, the steroid hormones never reach the hypothalamus where they are expected to exert their feedback inhibition, probably by inhibiting release of the releasing hormone. As a result, the ovary constantly receives follicular stimuli and follicular tumors develop. In fact, under the proper circumstances these benign tumors can be transformed into cancers.

Absence of sex hormone elaboration sometimes results from deletion of a sex chromosome, as in Turner's syndrome. In this disease the karyotype is missing one chromosome: it contains 44 autosomes plus one X chromosome instead of 46 chromosomes. Patients with Turner's syndrome have underdeveloped genitalia.

Glandular Hyperplasia

As was discussed in Chapter 4, several hormonal imbalances are caused by the development of tumors in the endocrine gland. Examples are acromegaly, associated with hypertrophy of the eosinophilic cells of the hypophysis; hypersecretion of corticosteroid hormones, associated with the hypertrophy of the adrenal glands; hyperparathyroidism, associated with hypertrophy of the parathyroid glands; and tumors of the islets of Langerhans, caused by hypersecretion of insulin with concomitant hypoglycemia.

PATHOLOGY OF CELL MEMBRANES

INTRODUCTION

The cell is something like a city. It is surrounded by a belt line (the plasma membrane) with specific spots reserved for entering or exiting the city. The belt line is linked to the administrative center of the city, the nucleus, through a complex maze of roads and streets. The administrative center may itself be surrounded by a parkway (the nuclear membrane). Between the maze of streets and roads are power plants (mitochondria), warehouses (storage sites such as those for glycogen and lipids), factories where new products (e.g., lipoproteins) are manufactured, and sanitation stations (detoxifying mechanisms, etc.). In the city food must be brought in, garbage and sewage must be removed, fabricated products must be transported from one part of the city to another or even be exported. Breaking the barriers of communication results in traffic jams, smuggling, and all sorts of catastrophies. Interruption of internal communications, for example, by a strike of public transportation or sanitation personnel, makes life uncomfortable, if not unbearable for everyone. Likewise, the cell is surrounded by a plasma membrane and contains a complex intracellular network of membranes delineating areas with more or less defined functions: the cytocavitary network, which includes the endoplasmic reticulum, the membrane that surrounds the area of cytoplasmic autodigestion (lysosomes), membranes surrounding the mitochondria, and membranes that surround the peroxisomes.

In this chapter we will focus on the plasma membrane and mention only briefly some of the changes that take place in the cytocavitary network.

More than any other cell structure except the nucleus, the cell membrane makes the cell unique. To consider the cell membrane as an inert envelope is to underestimate its function grossly. Many important decisions in the cell's life are made at the level of the cell membrane. In addition to preventing the exclusion of some basic constituents of the cell, the membrane determines the transport of components from the environment into the intracellular space and, vice versa, from the intracellular space into the environment. Enzymes and sources of energy within or close to the membrane determine active and passive transport. Many of the cell's social relationships are selected by the membrane, which may displace the cell by active movements on interphases or may establish contact with certain types of cells and repel others.

Because the membrane is at the confines of cell life, the cell has established feedback mechanisms between membrane and genome that regulate cell division and possibly other aspects of gene expression. In view of the importance of the cell membrane in cellular economy, it seems safe to predict that the elucidation in molecular terms of the structure and function of the cell membrane will have as much impact on cellular biology as the elucidation of the structure of DNA and the transfer of specificity from the genome to the finished protein. Therefore, much work has been devoted in recent years to the cell membrane, and several models for its structure have been proposed.

FUNCTION OF THE PLASMA MEMBRANE

Transport

The plasma membrane is a specialized cell structure that modulates the interaction between the cell interior and the surrounding environment and communicates with other cells. The cell membrane influences the composition of the cell by serving as a barrier and by regulating transport.

The cell membrane prevents the extrusion of cellular constituents in the medium and the penetration of some chemicals that are undesirable or unnecessary for the cell economy. But of more significance to cellular metabolism is the selective permeability of the membrane, permitting transport from the inside of the cell into the environment and vice versa. A great variety of molecules can be transported from the environment into the intracellular space through the cell membranes. They range from simple ones—such as water, ions, sugar, and amino acids—to extremely complex macromolecules—such as lipids (chylomicrons) and even other cells (as in phagocytosis).

The four major mechanisms of transport are: (1) free diffusion, depending primarily on the solubility properties of the compound to be transported; (2) passage through pores, a process essentially limited by the size and possibly the charge of the compound; (3) engulfment in expansions of the cell membrane (pinocytosis and phagocytosis); and (4) carrier-mediated transport.

The carrier concept assumes that the membrane contains segments of special molecular composition and configuration involved in transporting specific molecules. Carrier-mediated modes of transport are of two types: facilitated diffusion and active transport. The types of molecules that are transported and the mode of transport vary with the cell.

Because glucose, a water-soluble compound, is essential to the cell's nutrition, its transport is of particular significance. The hydrophobic lipid membrane impairs the penetration of glucose into the cell and therefore penetration of glucose requires special molecular mechanisms. The molecules involved are believed to be proteins called carriers because they transfer molecules from the outside to the inside of the cell. A carrier can act in two ways: it can carry the substrate across the membrane just as a ferry boat transports cars across a river, or it can function as a revolving door which by turning 180° transports the substrate from the outside of the cell to the inside. Glucose transport may occur either through facilitated diffusion or through active transport. Facilitated diffusion takes place in erythrocytes, lens, bone, and liver. The erythrocyte is the model for such a process. Penetration of the hexose is increased 10,000 times as compared to what one might expect from the lipid composition of the membrane.

Facilitated diffusion may be nonregulated or regulated by utilization of glucose. Nonregulated facilitated diffusion depends on the concentration of glucose in the extracellular fluid, the ability of the carrier to transport glucose from the extracellular to the intracellular aspect of the membrane, and the number of carrier molecules. Nonregulated facilitated diffusion is independent of the action of hormones. In erythrocytes the intracellular concentration of glucose is high. Glucose is dissimilated principally through anaerobic pathways. Glycolysis, the process of glucose breakdown by which energy is generated, is used principally for the maintenance of erythrocyte structures.

Regulated facilitated diffusion depends on the action of hormones or other activators. Characteristic of this mode of transport is the skeletal muscle. At rest the skeletal muscle depends on chemical energy, which is provided primarily by aerobic breakdown of glucose. But during exercise the energy generated by aerobic pathways cannot provide enough chemical energy for muscular contractility. Glycogen is broken down, glucose-1-phosphate is generated, and after conversion glucose-1-phosphate is dissimilated anaerobically through glycolysis. Soon the glycogen reserves are exhausted and more glucose is needed because the intracellular levels of glucose are very low. The penetration of glucose in the muscle cells is activated by a number of factors, including the rate of contraction, the presence of calcium, and, most importantly, insulin levels. The mode of action of insulin on the carrier is not known. In cardiac muscle the mode of transport of glucose is, with some variation, similar to that which occurs in the skeletal muscle.

Active transport requires energy for glucose penetration. It occurs in the epithelia—the absorptive epithelia of the intestine and the kidney tubules—because the metabolite must be transported into the cell against concentration gradients.

Transport is far from unidirectional. Secretion and excretion products are transported from the inside of the cell into the environment. Again, the cell may secrete molecules as simple as HCl or secrete complex enzymes or even entire organelles, such as zymogen granules. Secretion and intracellular transport are modulated by hormones—aldosterone for sodium, insulin for glucose, etc. Some of these hormones attach to the cell membrane to exert their function. Therefore, the membrane must have included in its structure some specialized segments that function as receptors for the attachment of the hormones.

Two cellular properties govern the relationship of one cell to another: movement and the ability to establish or reject contact with another cell. Cellular movement and communication are properties of the cell membranes. They play key roles in embryogenesis, morphogenesis, inflammation, wound healing, and cancer (invasion and metastasis).

The ultimate in transport cells may be the intestinal cells, which can absorb all components from small ions to gammaglobulin. At their luminal end, epithelial cells of the intestine have a special differentiation of their membranes. Long, slender, fingerlike projections called microvilli develop on the luminal border of the cell, and the assemblage of microvilli constitutes the brush border. The brush border is permeable to lipid-soluble and small water-soluble compounds, but many important nutrients—electrolytes, sugar, amino acids and special vitamins, gammaglobulin, and triglycerides—are brought into the cell by active transport. In most cases, specialized membrane structures are likely to be involved in the transport mechanism. In fact, in sugar transport, specific carrier systems have been identified.

Gammaglobulins, the proteins that confer immunity, are absorbed by pinocytosis. This process provides immune protection for a number of animals (pigs, cows, horses) which at birth are devoid of antibodies. A single feeding with colostrum brings the gammaglobulin level of the blood to that of the adult.

In view of these varied absorbent functions of the intestinal cell, it is not surprising that a number of enzymes have been found in the brush border. Some are hydrolases found in granules similar to those found in the pancreas, whereas others are hydrolases believed to be directly associated with the membrane structure.

Movement

To move, individual cells must either swim or crawl. Two types of crawling movements have been described; one has been studied primarily with amoebas, the other with fibroblasts.

Amoebas move by expanding pseudopods. After the cell has established a point of contact with a solid surface, that portion of the cell referred to as the tail contracts. The contraction projects a stream of cytoplasm in the opposite direction, distending a portion of the membrane into a pseudopod, which contacts the solid surface. The molecular structure of the "motor" that generates the movements is not known, but a number of hypotheses are available: (1) the tail contains contractible protein; (2) the peripheral cytoplasm is reversibly converted from sol to gel; (3) the elaboration of the new membrane provides the motive force.

It has long been known that fibroblasts and many other mammalian cells are capable of movement. Some of the details of the movement were revealed only when interference and surface contact microscopy became available.

The moving fibroblast adheres to the solid surface and forms a ruffled edge. The waving movement of the periphery of the cytoplasm, which is responsible for the development of the ruffled edge, secures displacement along the solid surface. The faster the cells move, the more rapidly their membranes develop the undulations that intermittently contact the solid surface. Again, the locomotive force is believed to be a contractible protein, forming short filaments inside of and parallel to the membrane. ATP is, of course, the source of chemical energy.

The direction of cell movement can be determined by three different mechanisms: contact guidance, chemotactism, and contact inhibition. In contact guidance, the solid substrate to which the cell must adhere for motion also determines the path of the cell because it is made of fibers (collagen) or grooves. Without such contact guidance, the movement of the cells is random, as occurs on smooth glass surfaces.

When a circular piece of amphibian skin is excised, the cells at the free edge start to move toward the center of the wound until they meet the cells that migrated from the opposite direction, then they stop. When two explants of fibroblasts are suspended in culture medium on a glass coverslip, the cells of each explant move at random as they proliferate, forming the ruffled edges described previously. The cells that move away from their respective explant toward the other explant come in contact between the two. As the cells make intimate contact at some point of their membrane, the ruffled edges disappear and the cells stop moving. This impairment of cell movement as a result of contact is called contact inhibition.

Contact inhibition is a rather specific event. Not every cell stops the locomotion of any other cell; for example, contact inhibition does not take place between fibroblasts and leukocytes or even between fibroblasts obtained from normal connective tissue and those obtained from sarcoma.

Obviously, directed cell migration coupled with specific contact inhibition could go far in explaining embryogenesis and histogenesis of tissues. The special cellular mosaic that composes a tissue may at least in part result from regulation of cell movement and contact inhibition.

Adhesion

Two opposing forces determine cell contacts: (1) a repulsion force that is electrostatic and develops because any particle suspended in an electrolyte solution is surrounded by double layers of ions, and (2) an attraction force due primarily to van der Waal's forces.

The repulsion forces decrease exponentially with the distance between particles, whereas the attraction forces decrease inversely with the distance. The resultant energy expressed in function of distance yields a sharp peak, the summit of which measures the potential barrier. On each side of this potential barrier are attractive troughs—one at short range and the other at long range.

Thus, two cells can make contact in two ways: they are within distances shorter than those at which the potential barrier develops, or the potential barrier is reduced.

The potential barrier can be reduced by increasing the attractive forces (little can be done about that) or by decreasing the repulsion forces. The latter can be achieved by altering the ionic milieu, reducing the dielectric constant of the medium, reducing the radius of curvature of approaching cell processes, and reducing surface potentials of approaching cells. The significance of all these parameters in determining cell contact remains unknown. When the potential energy has been overcome, adhesion between cells becomes possible. The forces that secure adhesion are not known. They could involve electrostatic covalents, hydrogen bonds, or van der Waal's forces. They may be direct from membrane to membrane, or indirect, involving protein or ions in the medium.

Observations made on mammalian cells have given some lead as to the mode of binding involved in cell adhesion. First, electron microscopy reveals that the points of contact are small relative to the cell size. Second, a protein must be involved in adhesion, since proteolytic enzymes facilitate separation. Third, exposure of the cell surface to neuraminidase also facilitates cell detachment, suggesting that sialic acid might be involved in cell adhesion.

Calcium's role in contacts of adult cells remains to be established. Calcium seems to affect the attachment of embryonic cells, and thus embryonic cells are readily separated by treatment with EDTA. In sponge cells, reaggregation is impaired if disulfide bond formation is blocked.

When embryonic cells are dissociated and then allowed to reaggregate, they do so by sorting out according to cell type. After the cells have segregated, groups of cells are distributed in the aggregate according to patterns similar to those found in the embryo. The selectivity of cell aggregation may be the consequence of qualitative specificity of the molecules involved in cellular adhesion, configuration of surface proteins, or quantitative differences in strength of adhesion.

What mechanism obtains in the completed organ has not been settled. Evidence for the existence of special adhesive substances has been assembled. For example, a compound apparently present in the cytosol of the cells of the living embryonic neural retina promotes histogenic aggregations of these cells. The compound is specific for retinal cells and will not aggregate liver, heart, or kidney cells. Little is known of the molecular structure of the compound.

Investigators have proposed that glycosyl transferases are involved in holding the cells together. If this were the case, the affinity between cells would depend on the molecular structure of the enzyme and that of the acceptor molecules on the membrane surface.

Because cellular aggregation is not thermodynamically reversible, investigation of the molecular events involved in cellular adhesions is complicated further. One can never be certain that all the bonds that make the cells adhere to each other are broken after pulling the cells apart. Regular reactions as we know them are usually thermodynamically reversible. Thus, if molecule A is reacted with molecule B to yield complex C, and the conditions are right, C can be split to restore the exact molecules A and B. However, evidence does not suggest that when cell A and cell B are made to adhere to each other, they can be again separated to restore cell A and cell B integrally.

This notion can best be grasped by considering a metaphor. If a carpenter glues together a piece of redwood and white pine and tries to separate the two pieces by shearing after the glue has dried, pieces of the redwood adhere to the white pine and vice versa. The two boards will be even harder to separate if the joints were dovetailed. The same is true of an attempt to separate two adhering cells. After separation, each of the cells retains some fragments of the other. The irreversibility of adhesion is best illustrated by experiments in which cells are shaken off their glass support. The cohesion of the glass molecules is such that no glass remains attached to the cell, but cell fragments remain attached to the glass and leave what have been referred to as "tracks" or "footprints."

The tracks contain mucopolysaccharides, proteins, and even RNA. What effects cellular fragments may have on their separated neighbors are not known, but fragments that include RNA and possibly DNA may affect the future of the segregated cell. That such cellular interactions are possible is suggested by electron microscopic observation. Cytoplasmic continuity between macrophages and lymphocytes in the lymph nodes and between the cells of peritoneal exudates of mice have been demonstrated by electron microscopy. Bridges containing DNA have been observed between tumor cells and peritoneal exudate cells of rats and mice.

Cell Communications

For a long time it was assumed that all cells were autonomous. Each cell was believed to occupy a specific geographic position in the tissue mosaic and be metabolically independent. Regulation of genome expression or rate of activity of bioenergetic pathways was thus considered the privilege of each cell. This belief collapsed when two simple but elegant experiments were performed. In one experiment an electric current was discharged in one cell and it propagated to other cells. In another, macromolecules were injected in one cell and they were transferred to other cells.

It was later established that the presence of calcium in the medium is indispensable for intracellular communication. Moreover, increasing the intracellular concentration of calcium to levels similar to those found in the extracellular fluids abolished intracellular communication. The cell membrane is believed to contain binding sites for calcium, and when all binding sites are occupied by calcium ions, the cell membrane is impermeable. When cells are in contact, the calcium attached to the binding site at the points of contact is in the presence of the intracellular media with low concentrations of calcium. Consequently, the calcium at the points of contact is released from its binding sites, making the junction permeable for intercellular communication. Although it is suspected that growth factors are transferred from cell to cell, it is not known what type of information passes through these intercellular communication paths.

Electron microscopists have described three major types of physical contacts between cells: desmosomes, tight junctions, and gap junctions.

Most of the epithelial cells—namely, those sheets of cells that form the skin and cover the surface of the gastrointestinal and respiratory tracts—are hooked together, not by a continuous glue molding the surface like the cement of a brick wall, but by joints that are specifically located at the membrane surface. These areas of physical contact between two cells have been known for a long time and are called desmosomes. Desmosomes were long considered to be sites of focal thickening of each membrane, but electron microscopy has revealed that this is not so. Thus, at the desmosomes two normal bilayer membranes are separated by a normal intermembrane space, which is filled with an electron-dense material probably proteic in nature. For unknown reasons, a similar electron-dense condensation is found at the internal aspects of the membrane at a point exactly opposite the membrane junction. The role of desmosomes is believed to be purely mechanical—they hold cells together.

Tight junctions involve the complete fusion of two cell membranes. Under the electron microscope the surfaces of the two cells are closely apposed at the site of the tight junction, and the classic appearance of the two electron-dense lines of the bilayered membrane has disappeared, as has the intercellular space. There are only three electron-dense lines, suggesting that the external layers of the bilayered cellular membranes have fused together.

The function of the tight junctions has been deduced from their anatomical location. They are indeed preponderant whenever a clear separation between two cell types or between a cell layer and the intracellular fluid is indispensable. For example, after intravenous administration, many substances never reach the brain, although they may reach all other tissues. The tight junction is believed to be the anatomical counterpart of the physiological barrier, the blood-brain barrier. Cells that line the blood vessels of the brain are held together by tight junctions; in contrast, such junctions are not seen in cells that line blood vessels anywhere else in the body.

The third type of junction, the gap junction, forms channels between two cell membranes, penetrating intercellular spaces and connecting the cytoplasm of each cell. The gap junctions are probably critical to communication from cell to cell, but their precise role is not known. However, the incidence of gap junctions varies with the developmental level of tissues: gap junctions are present in the blastula of some fish, but they disappear at later stages of embryogenesis. They are also believed to channel electrical communications between cells. These conclusions emerge from experiments with the electric organ of the electric eel. Rather than being adjacent to each other, the motor neurons that control that organ are linked together by gap junctions, which suggests that the junctions serve to secure synchronization of the electric discharge.

Regulation of Cellular Metabolism

Little is known of the cell membrane's role in regulating gene expression and the rate of cellular proliferation. Both in the embryo and in tissue culture, cell proliferation is slowed or stopped when cells make contact. The modulation of cell function by cell surface seems to take place through specific glycoproteins. Thus, glycoproteins regulate mitogenesis and serve as receptors for hormones, antibodies, viruses, and lectins.

The mechanism by which molecules and viruses interact with the surface receptors is not established, but two kinds of interactions—covalent and noncovalent—between surface binding agents and receptors have been described. Covalent interactions involve protein cleavage of the glycoprotein or specification of glycosyl transferases. Noncovalent interactions are of two types: those that involve only alteration of receptors, and those that involve global alteration of cell surface. In the first category are included antigen-antibody binding, viral attachment, attachment of cholera toxin, and hormonal binding; in the second are cross-linking by lectins and capping by a divalent antibody.

Continued synthesis of chondroitin sulfate in cartilaginous cells is dependent on cell apposition, and the interaction between cells stabilizes the cell surface and allows continued chondroitin sulfate synthesis. These findings suggest a feedback loop between the cell membranes and the genome.

The mechanism of pinocytosis has been investigated in *Amoeba chaos* fed paramecia. The inducer of pinocytosis leads to a structural change and an electrical change in membrane resistance. The electron-transparent core of the lamella of the unit membrane is at least twice as thick in areas of the membranes involved in phagocytosis as in others, and the electrical resistance of the membrane is reduced 50-fold in these areas. These morphological and electrical changes have been associated with the formation of typical tunnels and vacuoles that appear during pinocytosis. The formation of the pinocytic vacuoles depends on the concentration of external calcium. If the calcium in the medium is increased, the changes associated with pinocytosis are rapidly reversed.

With this information in mind, it is now easy to visualize how events occurring at the level of the cell membrane may influence cellular metabolism. We shall mention only three examples: regulation of genome expression, protein synthesis, and respiration.

We have seen that the sodium pump requires ATP for activity. ATP can be provided by aerobic or anaerobic bioenergetic pathways. Abolition of the aerobic pathways reduces the effectiveness of active transport mechanisms. In turn, active transport can influence the rate of respiration. Indeed, the cell membrane contains an ATPase that seems to be involved in sodium pumping by converting ATP to ADP. The ATP is provided from the intracellular bioenergetic pathways in which the mitochondria play the major role. The level of respiration in the mitochondria is itself regulated by the levels of ADP. The higher the levels of ADP, the higher the stimulation to respiration. Consequently, one may anticipate that when the sodium pump is activated, the levels of ADP will increase inside the cell, and the level of respiration will be affected. Such interrelationships between ion transport and cellular respiration have been demonstrated in brain and in kidney.

The transfer of amino acids from amino acyl tRNA to a polypeptide chain during protein synthesis requires optimal concentrations of potassium. Potassium depletion produces a block in the late stage of protein synthesis. When potassium levels are low, the cell continues to synthesize RNA but not protein. How the variation in potassium concentration regulates protein synthesis during development or during the steady state is not known, but potassium's effect on protein synthesis provides a new potential link between the cell membrane and regulation of intracellular metabolism.

STRUCTURE OF CELL MEMBRANES

Marchesi and his collaborators used a simplified method of extracting erythrocyte membranes—namely, the treatment of erythrocyte ghosts with lithium diiodosalicylate (LIS); all proteins are solubilized and can be separated by gel electrophoresis. Among a number of minor components, one protein, glycophorin, constitutes the major glycoprotein component of erythrocyte membranes. (Spectrin, another protein found in large amounts in erythrocyte membranes, represents approximately 40% of the total membrane protein and appears to be extrinsic to the lipid hydrophobic region.)

Glycophorin is made of 131 amino acid residues to which 16 oligosaccharides are attached and can be divided into three essential portions: the N-terminal half to which most of the sugars are attached, the C-terminal portion, which contains a cluster of charged amino acids, and the middle portion. The sugars form small oligosaccharide chains (e.g., tetrasaccharides), which may, for example, contain N-acetylgalactosamine, galactose, and sialic acid, but chains made of 8 to 12 monosaccharides are also found. The carboxy-terminal end contains charged amino acid; the middle portion of the molecule is hydrophobic.

What is peculiar to the glycophorin molecule is the clear dissociation of the hydrophobic and the charged residues. This arrangement provides for a liposoluble central strand held by two water-soluble ends, and, as we shall see, a new model for membrane structure has been proposed on the basis of the intramolecular distribution of charges in glycophorin.

To represent the erythrocyte membrane as a lipid layer into which a single protein, glycophorin, is dissolved would be inappropriate because the erythrocyte ghost contains several minor components with unknown functions and a number of specific enzymes. We shall return to this matter when we consider membrane models.

The study of the constituent proteins of membranes obtained from sources other than erythrocyte ghosts will not be outlined in detail here. Suffice it to state that: (1) often tissues composed of various cell types were used; (2) little effort was made to quantitate the yield of plasma membranes obtained, so the isolated preparation is not necessarily representative of the whole; (3) few investigators checked their preparation with appropriate morphological (electron microscopy) and biochemical (enzyme activities) markers. Even when markers were used, available techniques were limited. To prepare membranes from intact tissue, an investigator must homogenize cells, separate the cell components by differential centrifugation, and further concentrate the membrane preparation by separation on continuous or discontinuous gradients—for example, sucrose, dextrans, or ficolls.

When preparations obtained in this fashion are examined under the electron microscope, they appear as small vesicles believed to be fragments of plasma membranes that have curled up. Unfortunately, fragments of the endoplasmic reticulum or of the Golgi apparatus after homogenization yield similar vesicles; therefore, unless some special method is used to eliminate, or special criteria are used to assess, contamination of the plasma membrane preparation, one cannot be certain what one is working with. Histochemists have established that some enzymes (for example, the 3-diphosphoglycerate phosphatase in erythrocyte membranes or sialidase in the liver cell membranes) are primarily, if not exclusively, associated with the plasma membrane.

Even histochemistry has its limitations. The demonstration of an enzyme in one cell fraction by histochemical techniques may provide conclusive evidence for the presence of the enzyme in that fraction; however, it does not necessarily exclude the presence of latent enzyme, inaccessible to the substrate, in other fractions. Some activity of the enzyme used as a marker for the plasma membrane almost certainly will also be found in other cell fractions, particularly the endoplasmic reticulum where enzymes are synthesized. Thus, unless one has also available, in addition to the plasma membrane marker, a good marker for the endoplasmic reticulum (e.g., glucose-6-phosphatase in liver), it may be impossible to evaluate the level of endoplasmic reticulum contamination of the plasma membrane.

The enzymes associated with cell membranes can be classified in three functional groups: (1) those involved in transport, (2) those involved in secretion, and (3) those that affect the source of energy.

The enzymes associated with the plasma membrane have seldom been extensively purified and isolated. They have been identified only by their catalytic properties.

Relationships between protein and lipid in the cell membrane may not be only for structural arrangement. A number of enzymes have been shown to be activated by lipids. For example, phospholipase C is activated by phosphatidylcholine and lysophosphatidylproline.

MOLECULAR ORGANIZATION OF THE CELL MEMBRANE

Unless the cell membrane is made of a coded macromolecule similar to DNA, the multiplicity of its function would bring one to anticipate a rather complex structure. That DNA can be a macromolecule of relatively uncomplicated composition storing a triplet code is understandable because the function of DNA is relatively simple. It binds to a few enzymes, including DNA or RNA polymerase, and makes two different kinds of templates—its own complementary DNA chain and an RNA chain.

In contrast, the function of the cell membrane is varied. In fact, in addition to highly specialized functions, such as transport, many of the major cellular functions are represented in the membrane. It would seem that only the plasticity and the diversity of protein molecules could provide the membrane with such far-reaching potentials. Yet so little is known of the protein structure of the cell membrane that it is surprising so many attempts have been made to build membrane models and, even more astonishing, that some of the oldest models are likely to depict large segments of the membrane structure.

The study of myelin sheaths combined with physicochemical investigations of membrane permeability, electron microscopic examinations of cell membranes of various sources, and X-ray diffraction studies of some membranes led somewhat unexpectedly to the concept of the unit membrane. Today the two major hypotheses for the membrane structure are the unit membrane and subunit structure. We will examine them only briefly because a conclusive description of membrane structure will come about only when the molecular composition and the intermolecular reactions in the cell membrane are clearly understood. Although physicochemical, electron microscopic, and X-ray diffraction studies might provide a wealth of information to buttress one hypothesis or another, only laborious chemical breakdown of the cell membrane into its components will provide the definite answer.

Already in 1925 Gorter and Grendel proposed that cell membranes were made of lipid bilayers, but Danielli developed the concept further. Danielli proposed that the membrane is composed of a continuous bimolecular lipid layer in which the molecules are rigidly arranged. The hydrophobic bonds face inward, while the hydrophilic groups face outward. Consequently, polar groups are found at both sides of the membrane, and the lipid layers are believed to be covered with a continuous protein layer at their inner and outer sides, the protein being attached to the polar groups of the lipids. Phospholipids can play such a role in the bimolecular lipid layer.

The proportion of proteins to lipids satisfies the requirement of the model, and, as we have seen, the model is in keeping with the three-layered appearance of the membrane in electron micrographs.

If the chemical analyses are not absolutely convincing, they are not in conflict with Danielli's model.

Nevertheless, in recent years many objections have been raised against Danielli's model. Perhaps the most important concerns are its rigidity and its inability to explain the functions of the cell membrane. Studies with X-ray crystallography, infrared spectroscopy, nuclear magnetic resonance, and *in vitro* reconstructed models seem to indicate that the lipid layer, rather than being a typical bilayer, has a thickness less than twice the length of the phospholipid molecule. The fatty acid chains of the membrane interior are now believed to be in a disordered state approaching that of a liquid hydrocarbon. As it is described in Danielli's model, the membrane cannot provide for the variety of multimolecular interactions necessary for some of the membrane functions described previously.

The subunit theory proposes that the cell membrane is composed of self-assembled structural or functional subunits or both. In the cell membrane, the concept is purely hypothetical since such subunits have not been identified morphologically.

The difficulty that surrounds the demonstration of subunits in a given membrane is illustrated in mitochondria, in which two types of subunits have been identified. The first is a morphological subunit, often referred to as the inner membrane particle or the elementary particle, and it constitutes a repeating unit of the membrane that has been isolated and demonstrated to contain mitochondrial ATPase. The ATPase can be separated and rebound to the membrane. Whether the combination of the singular enzyme with the membrane qualifies as a subunit is debatable. The second type of subunit that has been described in mitochondria is a functional one known as the respiratory chain.

These findings have been challenged because some believe that the components separated are not pure lipids and proteins but also contain membrane fragments. If, indeed, the individual lipids and proteins of the respiratory chain remain attached to some membrane fragments, one may anticipate that these membrane fragments would spontaneously reaggregate without constituting a genuine subunit.

We will simply describe the position taken by Stoeckenius and Engelman. After exhaustively reviewing the subject, these investigators concluded that the membrane is likely to be made of a continuous lipid bilayer that provides the passive permeability required for cell survival. However, this bilayer is different from the traditional one. Instead of having a rigid arrangement of the bimolecular layers, the lipids form a disordered central hydrocarbon region. Proteins are believed to be located predominantly at the surface of the lipid layer to which they are connected by both hydrophobic and ionic bonds. Thus, the general permeability and the barrier property of the cell membrane are contributed by the lipid bilayer, whereas the selective permeability and function are contributed by special arrangement and conformation of the protein. Whether the specific properties of the membrane require that either lipids and proteins or proteins alone are arranged in a special mosaic to form a special subunit is not settled.

This theory of cell membrane structure reconciles some of the basic concepts of a unit membrane with the various functions of the membrane. Thus, all membranes would have a fundamental suprastructure: a bilayer of lipids coated with protein arranged in a disorderly fashion and detectable by electron microscopy. In addition, a much more diversified infrastructure is made of a combination of lipids and protein molecules. The infrastructure varies considerably from one side of the membrane to another, from one type of cell membrane to another, and within the cell from one type of organelle membrane to another.

This concept of a membrane composed of a fluid lipid bilayer containing floating proteins has been expanded into a more precise model thanks to the discovery of some special properties of cytochrome b_5 and glycophorin. The tripartite amphipatic (a central hydrophobic core and two hydrophilic ends) structure of glycophorin has already been described. Similarly, cytochrome b_5, a hemoprotein found in microsomal membranes, the polypeptide chain of which is composed of 100 amino acid residues, has its heme group protruding at the surface of the membrane (it can be clipped off with proteolytic enzymes). The amino acid sequence of the rest of the molecule indicates that it is highly hydrophobic, making the molecule well suited to mix with lipids. In fact, when cytochrome b_5 is separated from the membrane, lipids and then protein and lipids are again mixed together; the cytochrome instantly attaches to the membrane.

Rhodopsin is also believed to span the lipid bilayer of the retinal disc membrane. In addition, when the retina is exposed to light, calcium enters the membrane. This is believed to occur at the rhodopsin sites which form transmembrane channels similar to those formed by gramicidin.

On the basis of these findings, a new model for the relation between proteins and lipids in the membrane has been proposed in which the protein is believed to stretch from one end of the lipid layer to the other. Thus, the middle hydrophobic portion of glycophorin sinks into the lipid pool while its edges emerge from the pool. The inside end is loaded with the carbohydrate and the outside end with charged amino acids. This new concept is sometimes referred to as the iceberg model, comparing the protein with an iceberg swimming in a pool of lipids. The new model is entirely compatible with available electron microscopic pictures of membranes obtained by freeze etching, with the physicochemical status of lipids in the membrane, and with membrane function. Proteins found in the cell membrane are of the globular type. Globular proteins can have a bimodal structure often referred to as amphipatic (see Chapter 2). In the cell membrane, it is believed that the highly polar group protrudes from the membrane in the aqueous phase. In contrast, the nonpolar group is inserted in the hydrophobic interior of the membrane.

In the freeze-etching technique, membrane preparations are frozen in liquid nitrogen and cleaved under a vacuum, and the cleaved surfaces are sprayed with vaporized platinum and carbon. The vapors solidify and form a replica that can then be examined under the electron microscope.

Experiments with labeled ferritin, which attaches exclusively to the exterior membrane, have established that the cleaved surfaces never contain ferritin. This finding strongly suggests that the cleaved surfaces represent not the exterior but the internal surfaces obtained after fracture of the lipid bilayer. Studies with the erythrocyte membrane have shown that 10–20% of the total membrane is represented by globular particles 75 A in diameter and proteic in nature, which span the membrane width and penetrate both the inner and outer surfaces. This confirms predictions that the major glycoprotein component of the cell membrane spans the lipid bilayer and carries much of the carbohydrate and most of the sialic acid of the cell surface.

Among the properties that the composition of the lipid bilayer can be expected to modulate, one must include the physicochemical state of the membrane and the permeability of membranes to polar groups. Although the details of all the molecular interactions are far from known, the tightness with which two lipid molecules interact is reflected by the melting point of the molecular complex. Thus, the heating of lipid preparations brings about a switch from a state of "crystalline gel" to that of a "liquid crystal." The increment of temperature needed for passage from one state to another is called the transition temperature. By measuring transition temperatures of pure compounds and comparing them with those of bacterial or cell membrane preparations, one can determine whether the sample is in the crystal gel or the liquid crystal form. The existence of the latter form is, of course, of great significance because the liquid crystal state facilitates the passage of organic molecules that can dissolve in the membrane.

At least three factors help to determine whether the membrane or a portion of the membrane exists in one form or another: (1) the presence of cholesterol, (2) the length of the fatty acid chain, and (3) the degree of saturation of these chains.

The concept of a lipid fluid membrane in which proteins are dissolved is compatible with membrane function. For example, consider the carrier mechanism. The membrane, primarily composed of a phospholipid bilayer, is impermeable to substances from the exterior, and a carrier is responsible for transferring substances from the outside to the inside or vice versa. A carrier has been postulated to function by diffusing substances through the membrane and by operating as a revolving door.

The antibiotic valinomycin makes the membrane permeable to potassium ions but not to sodium ions. In other words, it modifies the membrane permeability in a selective fashion for a specific ion. Valinomycin is believed to achieve these roles by combining with the phospholipid layer membrane, then grabbing the potassium ion in the environment and diffusing it through the membrane. Once the potassium ion has reached the opposite side of the membrane, it is dissociated from the complex and released. This can be achieved because valinomycin is a nonpolar, fat-soluble polypeptide that can easily mix with the lipid bilayer of the membrane.

The model is also compatible with the notion of transport through channels. Thus, it has been proposed, for example, that some substances penetrate the membrane through channels that are lined with four constitutive membrane proteins that span the distance between the periphery and the inside of the cell.

Gramicidin A, a pentadecapeptide, facilitates ion transport through channels or pores. Although the details of the tertiary structure of gramicidin A are not known, the pentadecapeptide is believed to be twisted by formation of hydrogen bonds into a β-pleated helix. Two helical molecules of gramicidin A form head-to-head dimers, which are embedded in the lipid bilayer where they form channels that act as pores.

Immobilization of the lipid bilayer by cooling the cell in cultures at 32° C inhibits adenyl cyclase, indicating a close interaction between membrane lipids and proteins.

BIOSYNTHESIS OF THE CELL MEMBRANE

Since so little is known of the chemical composition of the cell membrane, one can hardly anticipate that the mechanism of cell membrane biosynthesis could be known, except of course for the elaboration of the building blocks—fatty acids for lipid and amino acids for protein synthesis. Even less is known about the metabolism of plasma membranes than about that of the endoplasmic reticulum.

A new plasma membrane must be elaborated after cell division, and the new membrane differentiates, thus developing specialized functional and morphological properties. Moreover, even in the steady state there is some protein turnover, and the rate of turnover varies from one protein to another.

The dynamic properties of the cell membrane can be better understood if one keeps in mind the performance of the cell membrane during the life span of some cells. At the onset of life the sperm cell, by mechanisms still unknown, pierces the membrane of the egg. This penetration creates a hole that must be repaired by fusion of the remaining membranes. Every time a cell divides, the single membrane that surrounds the two cytoplasms and nuclei must invaginate until it meets the membrane at the opposite side and fuse with it to allow separation of the two cells. In the process of pinocytosis, the membrane sends out evaginations that engulf the material to be pinocytized; but to secure the inclusion of the pinocytic vacuole, portions of the evaginated pockets must fuse. A similar but reverse process occurs in exocytosis; thus membrane fusion is a frequent and sometimes vital part of cell life. During embryogenesis, myeloblasts fuse to form muscle microfibrils. Some primitive cells fuse to form osteoclasts, and after conception cells are fused to form the syncytial trophoblast.

We have concentrated this discussion on cell membranes on the plasma membrane. However, it should be kept in mind that at least eight different types of membranes are found in the eukaryotic cell and even more are found in plants. These membranes outline special compartments in which metabolic constituents are present in different concentrations. Thus, they generate concentration gradients between the different compartments and the extracellular environment. The maintenance of these gradients is essential for the maintenance of life. To maintain the integrity of the cellular environment and the intracellular compartments, the cellular membranes must remain uninterrupted at all times (except for discontinuities compatible with selective permeability) and must, during cell division, maintain spatial and functional continuity.

When new membranes are synthesized, the existing membranes could be destroyed and be replaced by a new and enlarged membrane system. Such a mechanism would undoubtedly interrupt membrane continuity. Membranes could also be formed by expanding preexisting membranes simply by including new molecular components. Evidence from differentiating hepatocytes after induction of the endoplasmic reticulum with drugs suggests that the latter mode of membrane biosynthesis obtains, at least under those conditions.

One may also ask whether all the individual molecules that

constitute a membrane are introduced in the membrane multimolecular structure at the same or at different rates. Again, studies on differentiating hepatocytes of induction of endoplasmic reticulum after phenobarbital administration suggest that the expansion of the membrane is a multistep event.

THE CYTOCAVITARY NETWORK

Under the electron microscope the cytoplasm of eukaryotic cells contains spaces apparently filled with material of relatively low electron density. The cavities are separated by numerous membranes that are closely interconnected with each other and the plasma membrane, and that may represent areas of differentiation. This tridimensional complex of membranes and cavity is called the cytocavitary network. The principal membrane structures are the endoplasmic reticulum and the Golgi apparatus. In addition, a number of specialized granules are produced by the interaction of the Golgi and the endoplasmic reticulum. These specialized structures include pancreatic zymogen granules, whose role in digestion is clearly established; granules loaded with acid hydrolase called lysosomes, which in polymorphonuclears participate in the destruction of bacteria, but whose function is less obvious in other cells (except that they participate in the scavenging of the remains of dead cells); and the peroxisomes, whose role in metabolism remains a mystery.

The endoplasmic reticulum can itself be divided into two major classes of membranes, the rough and the smooth endoplasmic reticulum. The rough endoplasmic reticulum is built by polysomes and is the site of protein synthesis. The smooth endoplasmic reticulum is the site of synthesis of a number of other smaller molecules such as steroids and cholesterol, and through a special electron transport chain it also detoxifies drugs and harmful chemicals (not always successfully as we have seen in the case of carcinogens).

The origin of the cytoplasmic organelles is unknown, except maybe for the zymogen granules. The enzymes are synthesized in the rough endoplasmic reticulum and then transferred to the Golgi apparatus, where they are packaged and excreted from the cell by a process akin to reverse pinocytosis. The origin of lysosomes remains unknown except for the fact that the enzymes are synthesized in the endoplasmic reticulum. Some believe that the enzymes are then transferred to the Golgi and packaged to form the so-called primary lysosomes.

Little is known of the biosynthesis of mitochondria except that only a small portion of the mitochondrial protein is coded for by mitochondrial DNA. The other proteins are coded for by the nuclear genome and processed in the rough endoplasmic reticulum. How these proteins and lipids are assembled to form an organelle capable of generating ATP and of performing numerous metabolic reactions is still unknown. After this brief summary of the status of our knowledge of the cytocavitary system, one should not expect much more than descriptive alterations; no functional correlations of injuries of the cytocavitary system are yet available.

Endoplasmic Reticulum

We have already mentioned that carbon tetrachloride damages the endoplasmic reticulum, probably as a result of peroxidation, and interferes with protein synthesis. In the absence of proteins triglycerides do not complex with proteins to form lipoproteins, and fat droplets accumulate in the liver.

Barbiturates stimulate the formation of the smooth endoplasmic reticulum membranes. The cell is filled with them and they are thickly packed. When barbiturates are no longer administered, the pattern of the smooth endoplasmic reticulum membrane returns to normal. Some believe that this occurs discretely in an unknown way without involving focal cytoplasmic degradation. Others have claimed that the incidence of focal cytoplasmic degradation is markedly increased.

Benzoflavone inhibits the mixed-function oxidases, the detoxifying enzymes, and according to some reduces the carcinogenicity of certain chemicals that are transformed metabolically into active carcinogens.

Lysosomes

The role of lysosomes in cellular death is discussed in Chapter 9. Except for the fact that areas surrounded by membranes contain insoluble material such as silica, copper, mucopolysaccharides (in mucopolysaccharidoses), and complex lipids (in lipidoses), and that acid phosphatase (an enzyme associated with lysosomes) can be identified in these areas, the role of lysosomes in the pathogenesis of disease is unclear. The traditional view proposes that the exogenous insoluble substances are phagocytized by an invagination of the cell membrane which after fusion yields intracellular granules called phagosomes, which in turn fuse with primary lysosomes. The absence of digestion results from the insolubility of the compound ingested and the absence of the appropriate enzyme in the lysosome.

When the insoluble material is generated inside the cell, its accumulation is believed to be caused by the absence of a specific lysosomal enzyme. However, at present there are insufficient biochemical or electron microscopic data to support this hypothesis unequivocally.

A number of substances have been claimed either to labilize or stabilize lysosomes. The labilizers include hyperoxia, anoxia, X-irradiation, silica, vitamin A, bacterial endotoxins, streptolysin, carbon tetrachloride, and progesterone. Among the stabilizers are cortisone, cortisol, chloroquine, salicylate, and colchicine. The participation of these labilizers or stabilizers in cellular pathology is debatable for the following reasons: all studies are done *in vitro*, and there is not necessarily a correlation between the *in vitro* and *in vivo* findings.

In the case of X-irradiation, the labilizing effect *in vitro* could not be confirmed and certainly it could not be demonstrated *in vivo* until after cellular death had occurred. Similar results were obtained with carbon tetrachloride. Even the *in vitro* data are usually difficult to interpret because the labilizing effect is not linear with the dose of drugs added to the incubation mixture. In some cases low doses of the drug stabilize the lysosomes whereas larger doses labilize them.

Conclusive data on the role of stabilizers and labilizers of lysosomes could help elucidate their role in the pathogenesis of disease. Moreover, such data might help investigators determine whether leakage of intracellular enzymes does occur, and, if so, whether or not such leakage is deleterious to the cell.

Injection of lysosomal enzymes in joints can produce symptoms similar to those of rheumatoid arthritis, which suggests that lysosomal enzymes leak out of the cell when the cell dies. Nevertheless, it does not prove that the lysosomal enzymes leak out of the living cell. The interference of cortisone with the development of the symptoms seen in rheumatoid arthritis may be caused not by stabilization of lysosomal membranes but by prevention of cellular death by the hormone.

The agent causing tuberculosis, *Mycobacterium tuberculosis*, contains among other glycolipids multiacylated trehalose 2-sulfates. These sulfatides are believed to play a role in determining the virulence of the bacteria. Their exact mode of action is not clear, but it has been shown that they do interfere with the fusion of the membranes surrounding the phagosome (which contains the phagocytized bacteria) and the lysosome (which contains the digestive hydrolases) and thereby prevents the formation of phagolysosomes.

Peroxisomes

The participation of peroxisomes in cellular metabolism remains a complete mystery. These granules are rich in oxidative enzymes such as catalase and uricase. Catalase converts H_2O_2 into H_2O and may therefore play a critical role in protecting the intracellular constituents against peroxidation. However, patients with a hereditary deficiency of catalase (acatalasia) usually present only minor symptoms. There is no explanation for the extensive proliferation of peroxisomes that takes place after the administration of salicylates.

Mitochondria

Function and Structure Although the cell contains several enzymic systems generating energy, most cellular energy is provided by the intramitochondrial oxidation of the Krebs' cycle substrates linked to the phosphorylation of ADP in the electron transport chain. In addition to their role in oxidative phosphorylation, mitochondria are actively involved in ion transport.

With proper staining, mitochondria appear as granules or rodlike structures, both types existing in the same cell. Under direct illumination and by examining living cells with an oil immersion objective, an investigator views mitochondria as refractile filaments parallel to the long axis of the cell. Similar structures can be observed on examination of living cells with the phase-contrast microscope or after supravital staining with methylene blue or Janus green.

Electron micrographs of mitochondria demonstrate three main components: a mitochondrial membrane, a clear mitochondrial space, and internal membranes or septa.

In view of the osmotic properties of mitochondria, the structure and biochemical composition of the outer membrane is important. After proper fixation, the membrane is seen to be composed of three layers: two electron-dense layers enclosing an electron-transparent layer. The dense layers are thinner (50 Å) than the transparent (60–80 Å) layer. This triple layering is analogous to what was observed for the cell membrane. Sjöstrand has suggested that the clear central space is made of a lipid bimolecular layer, and the two outer layers are made of proteins.

Careful examination of this triple-layer membrane reveals that each dense layer can be resolved into two thinner layers enclosing a clear internal space.

Within the mitochondrial cavity, which is delimited by the mitochondrial outer membrane, are triple-layered septa. The total thickness of these septa is approximately 180 Å. They are composed of two dense layers 60-Å thick enclosing a clear interspace 60-Å thick.

Various relationships between the inner membrane in the mitochondrial cavity and the outer membrane delimiting the mitochondria have been described. Inside the mitochondria, the inner membrane may form independent septa or lamellae in a cavity that remains continuous. Another possibility is that the outer and inner membranes are in fact intimately connected. In that case, the septa may either form infolds of the three layers of the mitochondrial membrane or may just be projections of the inner electron-dense layer of the outer mitochondrial membrane. It becomes important to determine whether these projections end abruptly at some distance from the outer membrane or extend throughout the entire width of the mitochondrial cavity to form complete septa separating the main mitochondrial cavity into numerous smaller cavities.

The relationship between inner and outer membranes has been investigated by various methods, including osmotic change and electron microscopic studies. The osmotic behavior of mitochondria can best be explained by the existence of a bimolecular lipid layer, the internal monomolecular layer of which forms infolds. This interpretation implies that the cristae are infolds of the inner, denser layer of the mitochondrial membrane and could explain the shortening and disappearance of the cristae during swelling of isolated mitochondria. To increase its surface, the mitochondrial membrane would have to reduce the density of its infolds.

All mitochondria have internal membranes. Independent of the important roles that these internal membranes may play in mitochondrial physiology, either by subdividing the mitochondria into numerous small cavities or by increasing the reaction surface of the enzyme, they also constitute the morphological criteria for mitochondrial identification.

The disposition of the internal membrane is not alike in all mitochondria—the membranes vary in direction, number, and branching. Although they are frequently perpendicular to the long axis of the mitochondria, they are occasionally parallel to the long axis. Some mitochondria have few internal membranes; others, such as muscle mitochondria, have large and dense internal membranous components. Finally, the internal lamellae of some mitochondria, particularly some muscle mitochondria, form arborizations that give rise to reticular structures inside the mitochondrial cavities.

There seems to be no doubt that the cristae play an important functional role, for the density of the intramitochondrial lamellation correlates with physiological function, the lamellation being more abundant in tissues (skeletal muscle) with the greatest energy expenditures. Furthermore, there is a correlation between the number of cristae and the amount of cytochromes found in the mitochondria.

The picture obtained by electron microscopy is static, whereas the real mitochondrion is a dynamic body that bends or straightens its structure and can be displaced throughout the cytoplasmic field.

Motion pictures of mitochondria have even demonstrated mitochondrial fragmentation in both long and short axes and fusion of small mitochondria to form larger mitochondria. Mitochondria can travel from the cellular to the nuclear membrane where they appear to come in close contact with the nucleolus. The cause and significance of these movements remain unknown.

Inner and outer mitochondrial membranes are specialized modifications of a large family of membranes that includes the nuclear and the plasma membranes and the membranes of the endoplasmic reticulum. Mitochondrial membranes function as selective barriers that are also capable of active transport and provide a framework on which catalytic proteins (e.g., those of the respiratory chain) are tightly inserted in a pattern compatible with maximum efficiency. Like other membranes, mitochondrial membranes are composed of lipids and protein molecules—the fundamental building blocks. However, some features are unique to the mitochondria. For example, the mitochondrial membranes contain a lipid, cardiolipin, not found in the plasma membrane.

Mitochondria have not been excluded from the great membrane controversy in which the notion of a unit membrane is opposed to that of a membrane composed of subunits (see section on cell membranes). In fact, the concept of the subunit membrane emerged from studies on mitochondria. Clearly, the respiratory chain constitutes a functional unit, but whether it is an identifiable morphological unit is not certain. Separation and reaggregation of some portion of the respiratory chain as it has been described previously in Green's laboratory would seem to support the notion of a functional subunit, provided of course that none of the components used in the reaggregation are themselves membranous in nature. But it has been argued that the preparations used for the reaggregation do, in fact, contain membrane factors.

The inner particles, which at one point were believed to contain cytochromes, are now known to be devoid of respiratory enzymes and to contain a single enzyme—ATPase—which can be separated from and rebound to the membrane.

Although a great deal is known about the chemical composition of the mitochondrial membrane, and it is established that the membrane contains a number of catalytic proteins (e.g., the ATPase synthetase system, an ion transport molecular machinery and electron transport chain), the topological distribution of these proteins in the membrane is not known. All topological models proposed are at present hypothetical. However, it is accepted that the mitochondrial membrane, like most if not all biological membranes, is of the "fluid mosaic model" and is composed of a lipid bilayer traversed by proteins. Electron microscopic studies of the freeze-etch fracture faces of the outer and inner membrane indicate that the proteins are asymmetrically distributed not only when the inner is compared to the outer membrane, but also when the inner and outer faces of each of the fractured membranes are compared.

Origin After cell division, both daughter cells inherit mitochondria, and new mitochondria are formed before cell division occurs.

So much time was devoted to the study of the bioenergetic properties of mitochondria that investigation of their biosynthetic properties was neglected. But between 1960 and 1970 much evidence was assembled, which indicated that mitochondria possess an independent replicative system. Early morphological observation of live mitochondria with the phase microscope suggested that mitochondria divide. The presence of small amounts of RNA in association with mitochondrial fractions was considered to be in agreement with the autoduplication theory. Interest in autoduplication grew further when it was observed that mitochondria incorporate amino acids because this suggested that mitochondria were capable of synthesizing proteins. Work in the last decade has established that mitochondria possess the complete machinery for protein biosynthesis, including the templates DNA and RNA.

Mitochondrial DNA has a molecular weight of 9×10^6 to 10×10^6 daltons. The exact amount of DNA in mitochondria from various sources is not known because of the many pitfalls in measuring DNA content accurately; nevertheless, the amount of DNA in mitochondria is always small compared to that in nuclei. The amount of DNA in mitochondria varies depending on the source of the organelle. For example, although the DNA content of liver mitochondria is approximately 1.0–1.8 $\mu g/mg$ of protein, in rapidly growing tissues (mouse cell fibroblasts) it ranges from 2.5 (ascites tumor) to 5.5 (hepatoma) or even 8.2 (Walker carcinoma).

The base compositions of mitochondrial and nuclear DNA differ only slightly, but in some cases the differences are large enough that the two types of DNA can be separated by their bouyant densities. (Bouyant density is related to the guanosine-cytosine content of the DNA.) However, this is not the case for mammalian mitochondrial DNA, the buoyant density of which resembles that of nuclear DNA (nDNA) because of the small difference in base composition.

Although heat-denatured nDNA does not readily reanneal, heat-denatured mitochondrial DNA (mtDNA) does; therefore, the two types of DNA can be separated by sedimentation on cesium chloride gradients after denaturation.

A precursor is incorporated in mtDNA at rates different from its incorporation in nDNA. A typical example is liver: [³H]thymidine is not markedly incorporated in nDNA of normal liver, but it is in mtDNA. Thus, the specific activity of mtDNA is more than 40 times that of nDNA. In regenerating liver, the specific activity of mtDNA increases before the increase in specific activity of nDNA. Therefore, the metabolism of mtDNA seems to be independent of that of nDNA. Mitochondria contain a DNA polymerase with chromatographic properties different from those of the polymerase found in the cytoplasm. Moreover, incorporation of precursors into mtDNA results from *de novo* synthesis rather than from DNA repair.

Although there seems to be little doubt that mitochondrial DNA codes for protein synthesis, it is also clear that it does not code for all proteins found in mitochondria. It is not certain what protein mitochondrial DNA codes for. It has been proposed that it might dictate the biosynthesis of a structural protein, but not all mitochondrial proteins are coded by mtDNA.

A number of mitochondrial proteins—cytochrome c and glutamic dehydrogenase, malic dehydrogenase and leucyl tRNA synthetase, and all 53 structural proteins of mitochondrial ribosomes—are known to be synthesized in the endoplasmic reticulum and transferred to mitochondria.

On the other hand, the mitochondrial genome is believed to code for the largest polypeptide of cytochrome oxidase, an enzyme that has been highly purified and has been found to be composed of seven subunits. Mitochondrial DNA also codes for cytochrome b subunits. Similarly, there is evidence that the ATPase activity of *Saccharomyces cerevisiae* is under the control of the mitochondrial genome.

Even if the coding properties of mtDNA are limited, the biological implications are enormous. Not only does it shake the belief that all genetic control rests with nDNA, but it also allows for unexpected cellular metabolic individuality. Moreover, the finding of mitochondrial DNA raises the question of the existence of DNA with special coding properties in other parts of the cell, such as the cell membrane or the endoplasmic reticulum.

Thus, important questions emerge from these studies on mtDNA: Is the phenotypic appearance of a cell determined only by nuclear or by a combination of nuclear and cytoplasmic templates? Are some of the abnormal genetic traits transmitted from cell to cell by cytoplasmic genes? Both histochemists and biochemists have shown that liver and brain contain populations of mitochondria that differ from each other by their enzymic mosaic. As pointed out by Wagner, mitochondria may be heterogeneous simply because at all times liver and brain contain mitochondria at various stages of development. Yet it has not been excluded that heterogeneity is the consequence of "genetic" differences. If so, does the mother cell make sure that after cell division the two daughter cells receive an identical population of mitochondria, or is the cytoplasm of the two daughter cells heterogeneous? When female Mexican armadillos deliver their litter, they almost always deliver quadruplets. Even though the four embryos are derived from a single fertilized egg, marked individual differences are found in the biochemical mosaic of the offspring. Wagner asks whether these individual differences result from modification of the regulation of nuclear genes or from distribution of cytoplasmic templates among the dividing cells. Some of the morphological alterations that take place in mitochondria are listed in Table 8-2.

TABLE 8-1.
from Judy Berliner UCLA

Mitocondrial alterations	Cause
Swelling (drastic decrease in cristae area/unit volume)	Most types of cell injury, i.e., CCl_4 poisoning, alcoholism, necrosis in kidney
Enlargement (increase in size but not decrease in cristae number/unit volume)	Riboflavin deficiency, essential fatty acid deficiency, cortisone treatment

Presence of inclusion
    dark granules
    crystalline inclusions
	Lead poisoning,
	cadmium poisoning
	polymyositis in muscle

Increased numbers of cristae often completely filling mitochondria, usually defective in oxidative phosphorylation	Myopathies and Warthins tumors
Increased mitochondria mass/unit volume of cytoplasm	Hyperthyroidism

PATHOLOGY OF CELL MEMBRANES

Too little is known about the molecular structure of the cell membrane to be able to correlate molecular and functional events. Therefore, a list of pathological events will necessarily be incomplete. For clarity, we will distinguish between "structural" injuries and injuries caused by "functional" alterations.

Structural Injuries

Erythrocytes Many of the structural alterations in the erythocyte have been studied. Osmotic shock of the red cell can be caused slowly, for example by dialysis, or rapidly by plunging the red cells in a hypotonic solution. Three stages can be distinguished: the prelytic events, osmocytolysis, and hemoglobin exit. The most important alteration in the prelytic stage is the loss of potassium with entry of sodium and water, leading to progressive swelling of the cell. The cause of the potassium loss is not known, but it is certainly not linked to loss of cholesterol, phospholipids, or proteins because those events occur 24 to 48 hours after incubation.

In the rapid hemolytic process, the red cells swell and acquire a spherical shape 1 minute before lysis. Approximately 10 seconds before lysis, membrane defects—holes 2 A to 1 μm in diameter— can be observed in the shocked red cells. The holes tend to form in clusters. If lysis occurs slowly, hemoglobin oozes out of the cell in all directions, but if it occurs rapidly, the hemoglobin is ejected usually at one point of the red cell membrane. The membrane lesion is not irreversible, in fact, the membrane heals within 25 to 50 seconds of incubation in hypotonic solutions buffered with phosphate, but not in hypotonic solutions buffered with Tris. As a result of this restoration, ferritin that was able to enter the red cell ghost during the lytic period is unable to do so after repair.

However, in spite of the morphological sealing, the membrane continues to be leaky to ions for rather long periods of time, and only after further incubation (up to 1 or 2 hours) does the membrane become tightly sealed.

The erythrocyte can also be lysed by the action of detergents such as filimarisin (Filipin) or saponins. Those substances combine with cholesterol and form rings visible under standard electron micrographs. Freeze etching reveals that pits (which do not penetrate the entire membrane) are formed with filimarisin, whereas holes that perforate the membrane develop with saponins.

When sheep erythrocytes are incubated with sheep erythrocyte antibodies and complement, the reaction between antigen, antibody, and complement culminates in a lytic process that ultimately leads to hemolysis. Three morphological observations have been made at the surface of the erythrocyte subjected to immune lysis: (1) formation of rings (90 to 235 Å in diameter) at the extracellular surface of the erythrocyte membrane (the rings are not depressions because their center is level with the plane of the membrane); (2) aggregation of globules in the cleave-etched plane of the membrane suggesting a rearrangement of the protein component of the erythrocyte membrane (a correlation between the aggregation of globules and ring formation has been suggested); and (3) formation of small holes identical to those seen in osmotic hemolysis.

On the basis of these findings, investigators have proposed a sequence of steps for membrane lysis in the erythrocyte which involves binding of antibody to antigen, reaction with complement, prelytic loss of potassium with cellular swelling, hemolysis, and ultimately resealing of the cell membrane.

Drugs cause hemolysis of the red cell because of an enzyme defect like that of glucose-6-phosphate dehydrogenase, the presence of abnormal hemoglobins, or through immunological mechanisms. The molecular events that lead to the cellular lysis caused by drugs are not clear, but it is believed that drugs cause the prelytic loss of potassium and intrusion of sodium and water with further osmotic swelling and disruption of the cell. It has been claimed, but not confirmed, that drug injury requires ATP.

Congenital spherocytosis is a hereditary disease in which the red cell looses its normal disc shape to assume a spherical shape. The permeability of the membrane to sodium is increased and ultimately the swollen cell explodes releasing the hemoglobin. In at least some forms of hereditary spherocytosis the disease is believed to be caused by a mutation in the DNA coding for tubulin proteins.

Paroxysmal cold hemoglobinuria is a rare condition characterized by the sudden passage of blood in the urine after exposure to cold. The blood of affected patients contains a substance that reacts with the red cell membrane and lyses it, but the lysis takes place only after the blood has warmed. It is now established that the destruction of the red cell is mediated by complement, which reacts not with a specific antigen in the membrane but with a molecular defect in the membrane.

Erythrocyte ghosts treated with ATP and magnesium show ultrastructural changes of the surface membranes. These changes consist of internal vesicles resembling pinocytic vacuoles. The significance of these changes is still not clear. The permeability of the red cell membrane is altered also by doses of ultraviolet light one or two orders of magnitude greater than those needed to cause cellular death.

An example of membrane damage caused by molecular denaturation of the membrane is provided in the case of the aging red cell.

SDS-polyacrylamide gel electrophoresis of red cell membranes yields a number of proteins: a surface protein with a molecular weight of 90,000 to 110,000 daltons that can be attacked by trypsin, a protein with a subunit mass of 33,000 to 35,000 daltons which is known to be a D-glyceraldehyde-3-phosphate NAD oxidoreductase, and a high-protein subunit of 220,000 to 240,000 daltons. During storage of blood (a form of aging) the surface protein becomes denatured and can no longer be monomerized in 0.1% SDS. Monomerization requires SDS concentration ten times higher, yet since it can still be achieved with SDS it is obvious that the forces involved are not covalent and must instead be hydrophobic.

Another example of cellular death caused by alteration of the cell membrane is provided by the discovery of a nonspherocytic type of hemolytic anemia associated with an increase of phosphatidylcholine levels in the membrane probably as a result of defective membrane phosphatide catabolism. The half-life of the red cell is reduced to one-third of normal (approximately 10 days) and death results from a membrane defect. The red cell is incapable of protecting itself against ion and water influx, and the sodium pump, which is driven by ATP derived from glycolysis, becomes exhausted.

Other Cells Sometimes cellular death is believed to be caused by a direct attack on the membrane. Examples are the killing of the target cell by sensitized lymphocytes, or that of leukocytes by platelets, or of EDTA-treated blood in patients with Behçet's disease. In each case it is believed that the contact between the killer and the victim causes a tangential sheer force to be exerted by the killer on the target cell membrane. The sheer results in prelytic loss of potassium.

Although neither the electron microscopic details nor the biochemical alterations are known, it is possible that similar biochemical mechanisms obtain in the cellular death caused by ischemia in myocardial tissues. When the circumflex branch of the left coronary artery is occluded for 40 minutes, relaxation of the occlusion after that time will not prevent cellular damage. Yet during occlusion there are no changes in the electrolyte composition of the ischemic muscle. Reperfusion of the artery after 40 minutes rapidly increases potassium leakage with concomitant increases in water, sodium chloride, and calcium intake. Intracellular vacuoles appear and mitochondria swell. Dense bodies of calcium accumulate in the mitochondria. No significant changes in the lysosomes are seen. Whether the cellular swelling can be explained by the formation of holes in the membranes, as is the case in erythrocytes, remains to be established.

Thus, during the ischemic period anabolic events are slowed down, if not completely stopped, and catabolic processes cause alterations of the macromolecular structure of the cell membrane leading to leakage of potassium and penetration of water, sodium, and other electrolytes. The macromolecular alterations could result from the inability to replace a molecular component with a very high turnover, or from the denaturation of a molecule highly sensitive to ischemic conditions.

Polyene antibiotics (amphotericin B), mercurial and sulfhydryl groups damage cell membranes and cause leakage of cellular contents, principally electrolytes. The rate of leakage is usually greater than that caused by cardiac glycosides or ischemia. Despite the severe membrane damage caused by the agents described above, there is no increase in the incidence of focal cytoplasmic degradation nor rupture of autophagic vacuoles (lysosomes).

Streptozotocin and alloxan, two compounds known to cause experimental diabetes in mice when incubated with islet cells *in vitro*, cause a decrease in intramembrane particles of the islet cells' plasma membrane; the particles are normally detected by freeze fracture.

Injuries to Areas of Specialization of the Cell Membrane Some of the structural injuries that affect the cell membranes affect areas of specialization such as various types of junctions, microfilaments, microvilli, or even cilia.

Only a few examples of such types of injuries are presented. When microelectrodes are placed in liver cells the current is transferred from one cell to another. In liver cancer (hepatoma) the current is not transferred and the cell contacts such as the gap junction may not be seen. In advanced skin cancer the desmosomes, which link one cell to another, are often lost.

Nicotine paralyses the movement of the cilia of the respiratory cells. These cilia facilitate the evacuation of mucus and inhaled particles.

Spermatozoa are motile cells equipped with mobile tails. The tail is made of ciliary filaments linked by temporary cross bridges of the substance dynein, which is believed to generate movement in sperm tails. Dynein arms are also found in cilia and flagella. Patients have been observed with immobile sperm and lack of motility of the cilia of the respiratory tract resulting in tracheobronchitis and sinusitis. Electron microscopic studies of the cilia revealed that the patients lacked dynein arms.

In Chapter 2 we described neurons and their specialized cytoplasmic expansions, dendrites, and the synapse. Through the synapse neurons transmit signals to other neurons, muscles, or glands either by electrical transmission, probably through tight and gap junctions, or by neurotransmitters. In neurotransmission the presynaptic and postsynaptic receivers are separated by a gap. The neurotransmitters may either depolarize or hyperpolarize the receiver cell. The neurotransmitters, such as norepinephrine, serotonin, and acetylcholine, are contained in small vesicles that can be separated by differential centrifugation. To elicit their effect, the neurotransmitters must be discharged from these vesicles. The terminal part of the synapse can be separated by special techniques. These isolated components, called synaptosomes, are surrounded by a membrane that is an extension of the neuron plasma membrane. Within the membrane are mitochondria and synaptosome vesicles.

Among the neurotransmitters, acetylcholine is of particular significance. Acetylcholine modifies the permeability of some membranes to ions. This occurs by the interaction between acetylcholine and its specific receptor, a glycoprotein with a molecular weight of 250,000 daltons, probably a dimer composed of subunits (120,000 daltons) that span the width of the lipoprotein membrane.

A number of toxins affect the action of chemical neurotransmitters. Under some circumstances the content of the vesicle is not discharged. Botulism blocks the release of cholinergic granules. In contrast, the black widow spider toxin accelerates release of the toxin. Curare desensitizes the nerve ending to acetylcholine, possibly by interfering with receptors, whereas nerve gasses (cholinesterase inhibitors) prevent the breakdown of the neurotransmitter after it has exerted its effect. It has been suggested that schizophrenia and manic depression are conditions caused by synaptic disorders.

Myasthenia gravis is a synaptic disease that produces electrophysiological and structural anomalies at the neuromuscular junctions. The disease is frequently associated with autoimmune disorders, and thymectomy and corticosteroid therapy may be beneficial. Although antiacetylcholine receptor factors have been found in the gammaglobulin fraction of myasthenia serum, this does not mean that the disease is caused by antiacetylcholine receptors. A great variety of antireceptors have been found, and there seems to be little correlation between the symptoms and the concentration of receptors in blood.

Functional Injuries

A number of membrane lesions cannot be defined in molecular terms but can be identified by the distortion in function that they cause. They include a direct effect on enzyme, such as the Na^+-K^+ ATPase or adenyl cyclase alteration, damage to carriers, abnormal secretion, abnormal absorption, abnormal phagocytosis, disturbance of receptors, and disturbance of movements.

Interference with Enzymes Cardiac glycosides inhibit the Na^+-K^+ ATPase, a membrane enzyme that facilitates the extrusion of sodium and the intracellular penetration of potassium. The enzyme is also inhibited indirectly by massive drops of ATP in the cell as a result of ischemia.

The mercurial drug mersalyl inhibits adenyl cyclase of rat liver.

The inhibition is immediate and irreversible except by the addition of 2-mercaptoethanol. Similarly, the enzyme is inhibited *in vitro* by hexachlorophene. It is difficult to appreciate what effect these inhibitors may have on the enzyme *in vivo*.

Ca^{2+} ATPase is an ATP-hydrolyzing enzyme located in the membrane of muscle. The enzyme catalyzes the ATP-dependent transport from the membrane to the muscle cytoplasm (referred to as sarcoplasm).

Interference with Transport Cystinuria is a hereditary disease in which large amounts of cystine are spilled in the urine. When the urine is acid the cystine precipitates and forms stones, which interfere with kidney function. Along with cystine, ornithine, lysine, and argine are also excreted. Cystinuria is the prototype of a number of inborn errors of metabolism believed to result from deficiency of carrier proteins involved in transporting molecules through the cell membrane. Transport defects in kidney, intestine, or both have also been described for other amino acids—glycine, cysteine, tryptophan, and methionine—for glucose and galactose, and even for electrolytes such as calcium chloride and sodium. Similarly, phosphate diabetes, a disease in which phosphate is spilled in the urine, is believed to be caused by the absence of a carrier for phosphate in the renal tubule.

The red cell membrane contains a transport protein, probably a glyprotein, spanning the bilayer and having a molecular weight of 180,000 daltons. The protein is present in the membrane in the form of the dimer, which facilitates the anion exchange across the membrane. Sulfanilate and stilbene disulfonate bind covalently to the transport protein and inactivate it.

Interference with Secretion The lung is like a bagpipe; the central pipe, the trachea, branches off into two major bronchi, which then ramify into smaller bronchi. The small bronchi open into alveoli, polygonal structures lined by an epithelium made of two different types of cells, T1 and T2 cells. Alveoli are not completely occluded, but they communicate with each other through an opening on one side of the polygon. Before birth the lungs are collapsed. The alveolar epithelia are packed together, but soon after birth the alveoli expand. This expansion cannot take place without surfactant, a substance that reduces the superficial tension of the surface epithelium. The T2 cells of the alveolar epithelium start to secrete such a substance or group of substances during fetal life. In the fetus surfactant is secreted 20–30 days before term. The exact composition of surfactant is not clear. Crude preparations are complex and contain lipid proteins and mucopolysaccharides. The true lung surfactants are believed to be lipids that are retained in the air/water monolayer at the end of expiration when surface tension is extremely low. Several lipids are believed to be involved, but the principal surfactant agent is believed to be dipalmitoylphosphatidylcholine. The biosynthesis of dipalmitoylphosphatidylcholine starts with diglyceride, which in presence of cytidine diphosphate choline (Fig. 8-1) yields phosphatidylcholine plus cytidine monosphate. The phosphatidylcholine is deacylated by phospholipase A_2 to yield lysopalmitoylphosphatidylcholine in the presence of fatty acid. The addition of another fatty acid molecule yields dipalmitoylphosphatidylcholine.

The mode of breakdown of surfactant is not clear, but alveolar macrophages are believed to play a capital role.

Diseases of the lung appear when surfactant is either defective or in excess. Defective amounts of surfactant can be caused by absence or inhibition of synthesis or by accelerated breakdown.

Respiratory distress in the newborn, a major cause of neonatal death in developed countries, is believed to be associated with deficient surfactant production. Silicosis is believed to overstimulate the lung macrophage phospholipase A_2, which degrades phosphatidylcholine and reduces surfactant. Some have claimed, however, that silicosis is associated not only with fibrosis but also with the accumulation of lipids including cholesterol and dipalmitoylphosphatidylcholine. Alveolar proteinosis is believed to be caused by excessive amounts of dipalmitoylphosphatidylcholine, which does not exhibit surfactant properties. Table 8-1 gives a number of conditions in which surfactants are believed to be modified.

FIGURE 8-1. Biosynthesis of dipalmitoylphosphatidylcholine.

Diglyceride + cytidine diphosphate choline ⟶

phosphatidylcholine + cytidine monophosphate

| phospholipase A^2

lysophosphatidylcholine + fatty acid ⟶

lysopalmitoylphosphatidylcholine + fatty acid ⟶

dipalmitoylphosphatidylcholine

TABLE 8-1. Some causes of surfactant deficiency

Respiratory distress syndrome
Silicosis
Herbicide inhalation
Fluorocarbon inhalation
Cigarette smoking
Oxygen toxicity
Cardiopulmonary bypass
Pulmonary artery ligation
Forced hyperventilation*

* Assumed to accelerate the use of surfactant.

Defects or Injuries of Receptors The binding of two hormones principally involved in sugar transport, insulin and glucagon, is markedly reduced in some patients with juvenile diabetes and in obese insulin-resistant animals. This reduction in binding is believed to result from a decrease in the incidence of receptors for the hormone. For example, there is an inverse relationship between circulating insulin and the number of receptors. Whether the decrease in the number of receptors is the primary event, or whether the rise in circulating insulin causes the decrease in the number of receptors is not clear.

Familial hypercholesterolemia is a single-gene disorder causing elevated plasma concentration of low-density lipoprotein (LDL) with hypercholesterolemia. Because of the high blood lipid concentration, xanthomas develop in tendons and atherosclerosis appears at a premature age and causes coronary heart disease. The elevation in blood levels of cholesterol is believed to result from interference with the feedback inhibition of cholesterol synthesis. In normal cells the target enzyme for feedback control of cholesterol synthesis is 3-hydroxy-3-methylglutaryl coenzyme A reductase. Cultured fibroblasts of patients with familial hypercholesterolemia are defective in their ability to suppress the reductase. The regulator of the pathway is the plasma LDL, which is made of a complex of lipids (cholesterol, esterified cholesterol, phospholipids) and a protein, apoprotein B. The molecules are arranged to form an apolar central core made of esterified cholesterol and neutral lipids and a polar coat made of cholesterol, phospholipids, and apoprotein B. The cell contains special receptors for the LDL. Such receptors are missing in at least some patients with familial hypercholesterolemia. Consequently, the negative feedback of cholesterol synthesis exerted by LDL is inoperative.

Two types of receptors are specific for catecholamines: (1) α-adrenergic receptors are sensitive to epinephrine and norepinephrine but almost insensitive to isoproterenol (interaction of the hormone and the receptor is characterized by vasoconstriction); and (2) β-adrenergic receptors are most sensitive to isoproterenol and least to epinephrine (interaction between hormone and receptor causes vasodilatation and cardiac contractility). Little is known of the molecular structure of either type of receptor, but it is established that the β-receptors are not the same in all tissues and one distinguishes between β_1 and β_2 receptors (Table 8-2).

TABLE 8-2. Types of receptors in various tissues

Receptor	Tissue
β_1	Adipose
	Cardiac
β_2	Vascular
	Bronchial
	Uterine smooth muscle

In all tissues in which β receptors are found, the interaction between hormone and receptor stimulates adenyl cyclase. Cholera toxin enhances catecholamine-stimulated adenyl cyclase. It is, however, not certain that the toxin acts by binding to the receptors. Instead, it has been postulated that the toxin is incorporated in the phospholipid bilayer and complexes directly with adenyl cyclase. The toxin would then in some way mimic the effect of the hormone-activated receptor.

Similarly, the two types of histamine receptors are referred to as H_1 and H_2. The combination of histamine with H_1 receptor stimulates contraction of smooth muscles of the intestine and bronchi and is suppressed by the classic antihistamine drug mepyramine. The reaction of histamine with H_2 receptors stimulates gastric secretion and heart rate and inhibits contractions in the rat uterus. These reactions are inhibited by burimamide and metiamide but are insensitive to mepyramine.

Inflammatory cells, lymphocytes, polymorphonuclears, and monocytes contain chemoreceptor centers that allow them to respond to specific chemicals, which induce them to move toward the highest concentration of that substance. Such chemically directed movements are referred to as chemotactism. Substances eliciting chemotactism include products released by bacteria or dead cells. Obviously, chemotactism must involve membrane movements. The molecular events that cause chemotactism are complex. *In vitro* it seems that some components of complement activate a membrane protease. As may be expected, defects in complement components, especially in C3 and C5, may cause interference with chemotactism.

Inhibitors of chemotactism have been found in serum: a β-globulin (7S) directed against the C3 fragment of complement, and an α-globulin (4S) that inhibits the C5 fragment. The inhibitors are believed to act as proteases. The serum levels of these inhibitors are increased in Hodgkin's disease and in α_1-antitrypsin deficiency.

A chemotactic defect of unknown origin has been found in patients with the so-called Job syndrome, a disease that combines eczema, staphylococcal abscesses, and high serum levels of IgE.

Disruption of Inner Membrane Cytoskeleton The inner aspect of the cell membrane contains a cytoskeleton composed of two types of structures: microtubules and microfilaments. The microtubules are relatively large (15 nm in diameter) and rigid. They are found in the nucleus, where they participate in spindle formation, and in the cytoplasm, where they seem to be anchored to the cell membrane. Microtubules are polymers of a special protein tubulin. Microfilaments are thinner (6–8 nm in diameter) and form bundles or sheets, often apposed to the inner aspect of the plasma membrane. They contain contractile proteins, actin, myosin, and tropomyosin. Microtubules and microfilaments are believed to regulate cell shape, cell locomotion, cytoplasmic streaming, and mobility of cell surface receptors; moreover, they transmit signal inputs from the cell membrane to various outputs in the cell including the nucleus. For example, the capacity of lymphocytes to divide is impaired by alteration of the cytoskeleton. The formation of microtubules is energy dependent and is impaired by cyclic AMP and stimulated by cyclic GMP.

Glycophorin, spectrin, and possibly actin combine their properties to determine the red cell flexible conformation. Heating the red cell at temperatures (48–50°C) that degrade spectrin makes it swell and become spherical. Spherical and cup-shaped erythrocytes appear (and the condition is known as stomatocytosis) when the red cells are treated with vinblastine, a compound that tends to react with most fibrillar proteins. Hereditary spherocytosis is believed to be caused by a defect in spectrin, which interferes with its polymerization. The disease is associated with swelling, osmotic fragility, and potassium and sodium leakage.

A classic example of a substance that appears to bind to the cell membrane and thereby modify its macromolecular structure is phalloidin, a toxic cyclopeptide found in the toadstool *Amanita phalloides*. In some parts of the world, especially Europe, the collection of mushrooms is a sport. When properly prepared, mushrooms constitute a tasty dish whether they are poisonous or not.

Mushrooms are the reproductive organ or fruit of fungi. The fungus itself is mainly composed of filaments or hyphae that penetrate the soil or some other medium favorable for growth. Plant fungi may be saprophytic and grow on dead wood, parasitic and grow on the roots of living plants, or symbiotic. In the last case both plant and fungus benefit, often in unknown ways, from the cohabitation.

Several species of mushrooms that are indistinguishable, except by an expert mycologist, produce toxins. The species of the genera *Clitocybe* and *Inocybe* produce muscarine, which kills the victim in an hour or two. *Amanita muscaria* produces muscinol in addition to muscarine.

The species of the genus *Psilocybe* produce a hallucinogen, and *Amanita phalloides* produces two groups of toxins: the phalloidines and the amanitines. We will discuss the effect of phalloidines here; amanitines are discussed in the section on liver injury. Both phalloidines and amanitines are cyclopeptides. The first are made of 7, the second of 8 amino acids. The details of the structural requirements for toxicity are still incomplete. It is not certain that the cyclopeptide constitutes the integral toxin. Some investigators believe that the original toxin is a much larger protein molecule called myriaphalloisin or myriamanin. Although the injection of phalloidine into experimental animals is fatal, ingestion may not be. It is not certain that phalloidines cause death in humans, yet their biochemical effect at the level of the cell membrane is interesting.

In experimental animals, phalloidine is deadly. In a matter of hours it causes hemorrhagic necrosis of the liver. At autopsy, the striking finding is extreme swelling of the liver. Histologically, there is within 10 minutes marked vacuolization of the hepatocytes followed by necrosis. The mode of action of the toxin was investigated primarily in German laboratories. Tritium-labeled phalloidine was added to the perfusate of isolated livers and after differential centrifugation was found to bind to membrane fragments that sedimented with microsomes. When it became possible to obtain reasonably pure preparations of plasma membrane, it was shown that the labeled toxin preferentially binds to the plasma membrane. The toxin is believed to bind first to a glycoprotein receptor on the membrane surface. The receptor has not been purified yet.

In vivo and *in vitro* electron microscopic studies of the plasma membrane after injection of phalloidine reveal the formation of numerous microfilamentous structures at the inner surface of the membrane. The microfibrils at the inner aspects of the membrane are made of actin; this has been established by biochemical studies. If muscle actin is treated with phalloidine, it polymerizes to form F-actin filaments. Moreover, phalloidine binds to the actin and prevents its depolymerization.

It is not known how the polymerization of actin at the inner membranes causes the membrane distortion seen in phalloidine intoxication. Neither is it known how phalloidine penetrates the cell membrane to bind to the actin of the inner aspect of the membrane. In any event, the intoxication is followed *in vivo* by swelling and vacuolization of the hepatocyte. This vacuolization occurs *in vitro* by invagination, and *in vivo* by protrusion of the membrane. The difference between the *in vivo* and *in vitro* results has been attributed to changes in the cell environment. *In vivo*, the pressure of the intercellular fluid would force the membrane to invaginate, whereas *in vitro* it would be free to protrude. The membrane injury is further followed by elevated oxygen consumption of the damaged cells, loss of K^+ ions, and release of several enzymes.

Of course, as for most types of injuries, one must distinguish between a primary injury to the plasma membrane and its secondary effects. Plasma membrane of rat livers given carbon tetrachloride or phenobarbital showed reduced 5′-nucleotidase activity and activation of the Na^+-K^+ ATPase, two enzymes located in the plasma membrane. However, these events were detected only 24 hours after treatment and are likely to be several steps removed from the primary insult caused by carbon tetrachloride.

During phagocytosis, foreign material is engulfed in invaginations of the cell membrane, and the phagocytized material and its membrane envelopes are shed inside the cell. This process involves the loss of a large portion of the membrane surface. Yet there is no loss of transport and no new synthesis of carriers. Therefore, during phagocytosis carriers must be redistributed on the surface of the membrane preventing their disappearance during phagocytosis. This intramembrane movement is probably directed by microtubules because colchicine, which interferes with the polymerization of tubulin to form microtubules, prevents the movement of the carrier and is thereby responsible for loss in transport.

B lymphocytes carry IgG at their surface. After separation of the surface globulin, it is possible to prepare antibodies (antiglobulins) that will complex with the surface antibody of the lymphocyte, which acts as a receptor for the antiglobulin. When the antibody and the receptor molecules complex, they aggregate to form patches and the patches coalesce to form a single cap at the surface of the membrane. The formation of patches and caps implies that the receptors are mobile within the membrane. Patching and capping are inhibited by concanavalin A, a plant polypeptide. Concanavalin A is believed to prevent these events by acting on microtubules and microfilaments. Indeed, the interference of concanavalin A with patching and capping is prevented by colchicine. Moreover, concanavalin A also disrupts microfilaments.

Anesthetics related to cocaine, dibucaine, tetracaine, procaine, and xylocaine also inhibit capping. This effect is also believed to result from direct action of the anesthetic on the cytoskeleton. The effect of the anesthetic can be duplicated by the combined action of colchicine and cytocatalasin B, suggesting that both microfilaments and microtubules are involved. Moreover, electron microscopic studies reveal that the cells exposed to the anesthetic lose their normal shape to become rounded and lose their microtubules and microfilaments. It is believed that the disruption of the inner membrane cytoskeleton is caused by displacement of membrane Ca^{2+}. The anesthetics would thus increase the intracellular concentration of calcium to concentrations that favor the depolymerization of microtubules or affect calcium-sensitive proteins, such as actinomyosin, adenyl cyclase, or Ca^{2+} requiring ATPase.

Chediak-Higashi syndrome is an inherited disease found in animals including mice and humans. Two major symptoms dominate the clinical picture: absence of pigmentation (albinism) and defective polymorphonuclears with susceptibility to infections. Interference with the elaboration of microtubules is believed to constitute the basic molecular defect in the disease. Thus tubulin is formed but is not polymerized, possibly because the cyclic GMP–induced assembly of the microtubules is impaired either because of a lack of response to the stimulus induced by cyclic GMP or because of a defect in the elaboration of cyclic GMP. As may be anticipated, the interference with microtubule formation is associated with a defect in the mobility of the surface receptors in Chediak-Higashi disease.

Alteration of Cell Membrane in Cancer Cells Perhaps one of the most significant contributions to our understanding of the pathogenesis of cancer is the transformation of normal cells, cultivated *in vitro*, into cancer cells. The transformed cells exhibit many of the cytological changes observed in cancer cells *in vivo*, and when the transformed cells are injected into intact animals they cause cancer. Transformation was first achieved with viruses and later with chemical and physical carcinogens.

A great deal of the research done on cell membranes was inspired by the hope of understanding the pathogenesis of cancer. The cancer cell is free from contact inhibition of growth; it invades the surrounding tissues and metastasizes. These properties are likely to result from alteration of the cell membrane, although a conclusive explanation for the properties of the cancer cell is not yet available. There is definite evidence that the membranes of the cancer cell are altered. The alterations described include loss of surface protein, changes in the surface glycolipids, modifications of the surface degradative enzyme, increased receptor mobility, and development of new surface antigens.

In the early 1950s Miller's group at McArdle Institute in Wisconsin proposed the deletion theory of cancer. The postulate was that a critical protein was deleted in the cancer cell and that as a result the cell became independent of intracellular and extracellular controls. The missing protein could be a part of any portion of the cell, the genome, special organelles, or the cell membrane. A large cell surface glycoprotein (molecular weight 250,000 daltons) often referred to as LETS has been found to be missing in cells transformed *in vitro*. Moreover, the proteolytic digestion of the LETS protein transforms normal cells. The role of the LETS protein in cell physiology is not established, but it is thought to be involved in cell adhesion. Neither is it clear how the cell loses the LETS protein. It is believed that it is destroyed by a proteolytic enzyme, either plasmin or thrombin, which appears at the surface of the cancer cell.

Changes in the glycolipid composition of the cell surface have been demonstrated in a number of laboratories after viral transformation of cells. We shall consider only changes in ganglioside composition as an example.

Gangliosides are made of a ceramide moiety on which a polysaccharide side chain, as is the case for glycoproteins, is elaborated in a stepwise fashion by the successive actions of specific transferases. In transformed cells the activity of these specific transferases appears to be decreased or lost altogether. For example, a decrease in UDP-N-acetylgalactosamine hematoside-N-acetylgalactosamine transferase has been observed in SV40 and polyoma transformed viruses. In both cases, appropriate alterations in the glycolipid structure of the membrane are associated with the enzyme defect. Whether these changes are linked with malignancy is not certain. They could indeed be several steps removed from it and simply reflect changes in cell physiology accompanying transformation.

Neoplastic transformation *in vitro* is frequently, if not always, associated with surface membrane changes in glycosphingolipids. These changes result from a drop in the glycosyl transferase responsible for their synthesis. The glycolipids are, at least in part, synthesized in the Golgi apparatus, and many of the transferases are found in that organelle. Therefore, it is believed that the Golgi apparatus might play a role in the appearance of glycosphingolipids and glycolipids in normal and cancer cells and might indirectly regulate growth and possibly development of surface receptors.

Although the data are not always conclusive because the techniques are sometimes inadequate, it is generally believed that the activity of proteases and glycosidases is increased at the surface of cancer cells. The relevance of these observations to loss of contact inhibition of growth, metastasis, and invasion remains to be established.

Sugar and amino acid transport is generally increased in the cancer cell. The increase in transport can be produced in untransformed cells by treatment with proteases and glycosidases, or by addition of serum or hormones to the culture medium. In transformed cells there is a switch in the relative concentrations of cyclic AMP and cyclic GMP, in which the former decreases and the latter increases. The changes in cyclic nucleotide concentrations are believed to be responsible for the unleashing of uninterrupted cell division in transformed cells.

Antigens on transformed cells are more mobile than antigens on untransformed cells. After transformation antigens appear at the surface of the cell. Cells transformed by viruses present viral antigens. Cells transformed by chemicals present new antigens, but the antigens are unique for each transformed clone. Antigens present in the fetus but not in the adult appear in many transformed cells.

The presence of new antigens on the surface of the transformed cells, in cancers produced experimentally and in human cancers, has generated new hopes for improved diagnosis and possible control of cancer by appropriate immunotherapy. The future of immunology in cancer diagnosis and therapy is difficult to predict. Certainly, some important contributions have been made in both fields. The discovery of the presence of a carcinoembryonic antigen in the blood of patients suffering from cancer of the intestine has helped not only in diagnosing the disease but also in following the efficacy of therapy. Thus, the levels of carcinoembryonic antigens, which may be high before therapy, decrease after surgery or even after administration of the appropriate chemotherapeutic agents.

Although there have been some reports of impairment of the growth of cancer by stimulation of cellular immunity, cancer cells often bypass immunological defenses in various ways: by shedding the antigen, masking it, or redistributing it so that the host defenses are unable to detect the antigen. Even if host defenses detect the antigen, the antibodies cannot destroy the cells.

To kill a cell the antigen must first complex with the antibody and then the antigen-antibody complex must fix complement. Complement fixation requires a specific distribution and configuration of the adjacent antibody molecules bound to the surface antigen. In cancer cells the surface antigens are more mobile than in normal cells, and it is conceivable that the ease with which the antigens are redistributed causes the cancer cells to escape complement fixation and killing. The cancer cells readily shed antigens or antigen-antibody complexes. The shedded material provides so-called blocking factors that bind to the cells destined to attack the cancer cells and thereby neutralize them.

At present it is impossible to predict how much immunity will contribute to control of the growth of cancer cells because it appears that in the stepwise alteration of gene expression which seems to be characteristic of a population of cancer cells, some cells sooner or later manage to develop properties that permit them to escape the immunological defenses. Still, it cannot be excluded that a combination of existing forms of therapy and immunotherapy might help to cure at least some cancers.

CONCLUSION

The study of the cell membrane is at the frontier of modern biology. Because of the difficulties in preparing membrane components in pure form and the difficulties in reconstituting the original membrane, progress has been slow. Yet in the last decades great progress has been made in understanding the structure and function of the cell membrane, which has made it possible to describe meaningful membrane models and to interpret the mode of action of certain drugs and the pathogenesis of certain diseases.

REFLECTIONS ON CELLULAR DEATH

INTRODUCTION

Because of fear of one's own death, or because of the grief suffered by those who remain alive, the quick and the dead are at antipodes in the minds of men. Yet life and death continuously interact in the living world. Today's foods are packaged in a variety of containers and their color, shape, and sometimes taste are far removed from those of the original product. Thus, it is easy to forget that man kills cows that eat grass in a green pasture, or that salmons swim upstream to the Kodiak shallow lakes to spawn and ultimately become the prey of bears, eagles, and foxes. Animal life thrives on other forms of life.

Even the harmony of the individual living organism depends on the delicate balance between cellular proliferation and cellular death. To live we must die a little all the time.

It is likely that most cells, even those with long life spans, die as a result either of turnover of cellular organelles or of discrete macromolecular destruction.

Nature has built-in cycles of life and death. Outside of the tropics, every autumn almost every leaf dies slowly, and it is no wonder that men refer to old age as the winter of their life. Growth and maintenance of the animal organism result from the delicate balance of cellular proliferation, maturation, and elimination. During embryogenesis millions of cells are killed or commit suicide in preselected areas. After elimination, the dead cells are replaced by other cells, often of a different type.

Morphogenesis of bone tissue results from repeated destruction of preexisting histological structures with replacement by new ones. Red cells, polymorphonuclears, and lymphocytes of the blood all have limited life spans, yet the total circulating number of these cells remains constant. Superficial layers of skin and mucosae are shed periodically, but never are surfaces denuded in the healthy organism. These few examples suggest that cellular death is as important to organismal health and survival as cellular proliferation. Moreover, it is likely that the life of cells depends on balanced elimination and replacement of intracellular organelles. Consequently, in normal cells catabolic and anabolic systems exist side by side.

Because of this unexpected but indissoluble marriage between life and death, the cells—which have, since Virchow, been considered the unit of life and the target of injury—can preserve their harmony only by delicately balancing anabolism and catabolism.

Such balance can be maintained in at least two ways: (1) regulation of substrate and enzyme resulting in an equilibrium that is at some time tilted toward anabolism and at others toward catabolism, and (2) maintenance of catabolic enzymes in the latent state (see Chap. 3).

PRIMARY, SECONDARY, AND CRITICAL INJURIES

When the environment of the cell is modified so suddenly or so drastically that smooth adjustment of the intracellular events to the insult is impossible, the cell is injured. With the possible exception of fixation, few insults stop all cellular reactions at once. Most injuries are compatible with at least temporary persistence of many if not most intracellular functions. However, as a result of injury, membrane permeabilities are modified, bioenergetic pathways are rendered ineffective, mitosis is blocked, lipid or other intra- or extracellular material accumulates, and the delicate balance between catabolic and anabolic pathways is tilted one way or the other.

Because the cell presents such changes after injury, it is often said to react to injury. The expression "reaction to injury" implies that the changes that take place in the injured cell are not passive consequences of the insult but rather are dynamic responses to it, either to secure survival or to accelerate death. Responses to injury can be primary or secondary. Special injurious agents have affinities for specific molecules. For example, cyanide blocks the activity of cytochrome oxidase, and many antimetabolites interfere with the active site of specific enzymes, thereby preventing metabolism of normal substrates. Biological agents secrete cytotoxic substances and cause immune reaction, which kills the cells that harbor them. These are primary injuries.

The primary injury is not always restricted to a single molecule—it may damage several molecular systems at the same time. For example, although there seems to be no doubt that much of the damage produced by ultraviolet light results from the formation of thymine dimers in the DNA molecule, ultraviolet light might also damage other macromolecular systems. Similarly, X-irradiation of bacteria or mammalian tissues interferes with DNA synthesis. This interference is likely to result from insult to the DNA molecule. Yet because of the nature of X-irradiation, it seems unlikely that the primary injury affects only DNA. Nevertheless, it is quite possible that damage to the DNA molecule is the only event critical to cell survival.

In spite of the need for distinguishing between primary and secondary and between critical and noncritical injuries, pathologists use the phrase "reaction to injury" to refer to intracellular changes that take place after the insult. Yet in some cases it is obvious that the changes are an inevitable consequence of the primary molecular injury. The accumulation of glycogen in glycogenosis or of lipids after carbon tetrachloride intoxication is simply the inevitable consequence of an insult to other molecules.

The primary biochemical insult is usually followed by a cascade of secondary metabolic alterations. A case in point is the mutation in the DNA molecules that is responsible for absence of uridyl transferase and thereby leads to galactosemia, which is believed to result from the accumulation of galactose-1-phosphate, an inhibitor of a number of enzymes. Inhibition of these enzymes could explain the development of hypoglycemia, jaundice, and cataracts, and possibly the mental retardation that develops in galactosemic patients.

PROGRAMMED DEATH

To understand elementary mechanisms in cellular death, it is useful to distinguish between programmed (necrobiosis) and provoked death (necrosis). The fate of the red cell best illustrates the programming of cellular death. In a well-regulated sequence of steps the genome is systematically repressed for the purpose of making a single protein, hemoglobin. The human red cell ultimately sacrifices the opportunity for reproduction and long life span for the sake of efficiency by discarding its useless nucleus. It is, however, not known whether it commits suicide or is sentenced to death by agents foreign to its intracellular environment. The fate of the epithelial cell of the skin which ultimately becomes stuffed with keratin is not very different from that of the red cell.

If we know little of what causes programmed death in the red cell, we know even less of what triggers it in the fetus, where programmed elimination of cells of various types is indispensable for development. Evidence from plants in which programmed death is rigidly regulated can be helpful, but even in plants we know little about the mechanism that triggers death.

If the harmonious deployment of cellular proliferation, cellular maturation, and cellular death were never disturbed by changes in the environment, the living being could be eternal or at least live until an inherent death clock programmed to end life was put in motion.

But elementary molecules of life—such as nucleic acids, which probably required UV light to develop—were threatened long ago by the very light that caused them to be born. As life evolved further, living organisms collided with their environment physically and died from trauma. They inhaled or ingested poisons and were attacked by other living organisms, often invisible ones. As these few examples illustrate, the rate of organismal and cellular death is greatly accelerated over what it could be without interference from the environment.

It is not known at present whether injuries accelerate a death clock or whether they trigger cellular death by mechanisms other than those that obtain under physiological conditions.

Necrobiosis, insect metamorphosis, and chick embryo development clearly illustrate the programming of cell death. Some insects, for example, moths, butterflies, and mosquitoes, develop in three stages. The caterpillar is a larva, which after molting several times becomes a pupa. The amazing metamorphosis of a caterpillar into a butterfly takes place during the pupal stage. Various appendages, especially limbs and wings, originate from internal buds, which by a combination of cellular death and proliferation are ultimately molded into the adult shape. The salivary gland of the larva undergoes histolysis 10 hours after the larva enters the pupal stage. The cytonecrosis that accompanies histolysis is initiated by a hormone that induces metamorphosis. If the salivary glands of the larva have not entered the pupal stage and are transplanted to a metamorphosed larva, the graft undergoes hydrolysis. However, the stage of development of the larva from which the salivary gland is obtained is important because the salivary gland cells can respond to the hormone only after a certain stage of development, the second molt. Salivary glands obtained from a larva that has reached only the first molt transplanted to a metamorphosing larva will not undergo hydrolysis.

Spinal ganglia from the early stages of chick embryos seem to contain more neurons than are needed for adequate innervation of most of the adult body, except for those parts of the chick's body that include the wings and the limbs. Although a large number of cells die in the ventral portion of the cervical and pelvic ganglia, few of the cells die in the brachial and lumbar ganglia. However, the cells of the lumbar and cervical ganglia die if the limbs are amputated. All these observations and many others clearly demonstrate that cells do not die by chance during necrobiosis, but that cell death is rigidly programmed in space and time. Some unknown intracellular or extracellular factors make certain cells competent to die in due time. From that moment they become competent cells, following a rigid course timed by a death clock. Unfortunately, nothing is known of the molecular events that take place in the setting and unwinding of the death clock.

PROVOKED DEATH

Various types of injuries have been invoked as a cause of cellular death. Such injuries include interference with the expression of the genome, reduction of the source of chemical energy, alterations of the cell membrane, and activation of catabolic pathways.

Interference with DNA synthesis, such as that associated with the administration of ultraviolet light, X-irradiation, or alkylating agents, has been associated with cellular death. Similarly, blocking of mitosis with substances such as colchicine and vinblastine, which interfere with the assemblage of the tubular structure needed to elaborate the mitotic apparatus, is known to cause cellular death. Inhibition of the enzymes involved in DNA synthesis causes cellular death, at least under some circumstances.

Actinomycin D, which intercalates between the two strands of DNA and interferes with transcription, can also cause cellular death. Similarly, in bacteria mitomycin, rifamycin, and antibiotics that interfere with transcription cause cellular death.

Molecular Mechanisms in Provoked Death

The mechanism initiating cellular death (interference with DNA replication or transcription) is unknown. For a long time it was suspected that cellular death resulted from loss of the ability to produce ATP through either glycolysis or aerobic pathways. All evidence suggests that such a mechanism causes death only in special circumstances.

Damage to the cell membrane through osmotic shock is obviously a direct cause of cellular death. Similarly, damage caused by immunological responses is sometimes the primary cause of cellular death. However, in many other cases it is difficult to determine whether the cell membrane alterations seen during cellular death are primary or secondary to other forms of intracellular damage.

In all cases of cellular death lesions to macromolecular structures are certainly involved, but whether the lesions themselves cause death is not certain. It cannot be excluded that depending on the type of cell involved and the type of damage incurred by the cell, a death clock is put in gear.

The Point of No Return

Many types of injuries cause cellular death, which may ultimately result in bodily death or serious mutilation. To prevent injury is often impossible; therefore, the only hope for preventing the consequences of injury is to reverse what seems to be an inevitable course. Although we know little of the chain of events in cellular death, it seems obvious that death is not instantaneous but is associated with a sequence of molecular and ultrastructural changes. The earliest changes probably are reversible: the latest are not. Before the changes can be reversed, we must determine which event separates the last reversible from the first irreversible step, and how soon these critical events occur after injury. The time at which this critical molecular change takes place is the point of no return.

If diagnosis is prompt, knowledge of what happened at the critical moment should permit physicians to reverse the sequence of cellular changes that could otherwise inevitably lead to cytonecrosis. No experiment has conclusively established where the point of no return lies, and it is likely to differ depending on the cell type. A great deal of research in that field has been done with ischemic cells, but the experiments are too complex to be described here. Suffice it to point out that cells can survive with a low level of ATP, but recovery of the ability to synthesize ATP appears to be essential for cell survival. It is conceivable that cells are programmed to function at levels of ATP-regenerating capacity far above those actually needed for adequate performance. It could be further assumed that as soon as the ability of the cells to maintain ATP-regenerating capacity has dropped to a certain threshold, the death clock is set and cytonecrosis is inevitable. β-Galactosamine causes necrosis in the liver, and a certain level of deprivation of the substrate UTP appears to be critical to this process. If deprivation is maintained beyond a certain period of time, the lesion is irreversible. Whether or not the triggering of the death clock is linked to protein synthesis remains to be established, but there have been claims that inhibition of protein synthesis with cycloheximide protects against radiation damage.

Catabolic Enzymes and Cellular Death

One would naturally expect catabolic enzymes to play a role in cellular death.

Three groups of diseases are probably caused directly or indirectly by disturbances of the mechanism activating proteolytic enzymes. The activation of trypsin plays a key role in the pathogenesis of acute pancreatitis, and the absence of α-antitrypsin is responsible for some forms of emphysema and for some cases of cirrhosis of the liver in children. The pathogenic mechanism involved in each of these diseases remains unclear.

Although in the last decades much of the work on the role of catabolic enzymes in cellular death focused on acid hydrolases, the existence of numerous other catabolic enzymes cannot be ignored in the study of cellular death. The role of catabolic enzymes in cellular injury usually is unknown, but in a few rare cases it has been shown that their role is complex and that they may repair injuries or cause further damage (see Chap. 5).

DNA, a child of the sun, was also among its first victims—as mentioned previously, UV light causes among other injuries the formation of thymine dimers. Moreover, iatrogenic or environmental chemicals bind to the deoxypolynucleotide *in vivo*. Much of the damage to DNA is repaired by the sequential action of an endonuclease, alkaline phosphatase, and DNA polymerase. But in some cases, for example, after the administration of X-irradiation, the repair mechanism fails and is likely to cause double-strand breaks for which no repair mechanism is known, at least in mammalian cells (see Chap. 5). It is likely that a battery of catabolic enzymes is involved in DNA repair, and it cannot be excluded that many enzymes still to be discovered participate in the repair of RNA, proteins, and other macromolecules.

In the latter part of the 1950s, de Duve and his associates in

Louvain isolated a cytoplasmic organelle rich in acid hydrolases. The enzymes were most effective at pH 5 and exhibited latency, in that full activity could only be released by blending and osmotic shock or by the action of detergents. The granule containing the acid hydrolases was named the lysosome. The hydrolases were found to be synthesized in the endoplasmic reticulum. In the traditional view they are transferred from there to the Golgi apparatus, where they are packed in an envelope. These small granules containing the hydrolase are the primary lysosomes.

Primary lysosomes are called on to perform intracellular digestion whenever they are needed. Their movement at the site of digestion is believed to be directed by microtubules. At the site of digestion they combine by fusion of membranes with the portion of condemned cytoplasm. The enzymes are released and the cytoplasmic content is progressively digested. During digestion the cytoplasm acquires different electron microscopic appearances. The combination of the primary lysosome with the doomed portion of the cytoplasm forms the secondary lysosome.

The lysosome is the only cellular organelle that was identified biochemically before it was seen morphologically. de Duve quite logically directly linked the granule with cellular death by proposing what was then an appealing hypothesis, that of the "suicide bag." The suicide bag theory postulated that cellular death was triggered by the release of the lysosomal enzymes into the medium.

In 1938 it was recognized that cellular death is caused by a chain of events: a decrease in oxidizing metabolism resulting from subnormal oxygen tension in the tissues; an increase in concentration of hydrogen ions resulting from the accumulation of acids including lactic acid, CO_2, and other metabolites; and an increase in proteolytic activity. The mechanism of activation of proteases was not obvious, but it had been shown that in tissues all cathepsins do not exist in the free state—some remain permanently bound to insoluble proteins after attempts at extraction.

Since 1938 a number of studies on injuries associated with cell death have shown that a variety of molecular alterations take place minutes, or sometimes hours, before the release of acid hydrolase.

In pancreatic autolysis, release of the enzyme from zymogen granules precedes that of acid hydrolases from lysosomes. In liver autolysis, acid hydrolases are released when the levels of ATP have fallen to zero. The release of acid hydrolases starts 15 minutes after the onset of autolysis. After 3 minutes of autolysis the levels of ATP have fallen to 5% of their normal intracellular concentration. At 10 minutes they are down to zero. It is also known through the work of Judah and his associates that severe mitochondrial alterations take place between 5 and 30 minutes after the onset of autolysis.

In ischemia of the kidney most of the critical events concerning lactic acid formation and ATP generation occur in the first minutes after the onset of autolysis. The release of acid hydrolase does not start in kidney until 30 minutes after the onset of ischemia and reaches a peak only 1 hour later, again indicating that the release of such hydrolases follows the development of other macromolecular alterations and is likely to be a consequence of the injury rather than the primary manifestation of it. Studies on liver damage produced by hypoxia also support the notion that the effect of lysosomal enzymes in cellular death is several steps removed from the original insult.

X-Irradiation induces almost instantaneously single-strand breaks, double-strand breaks, cross-links, and base alterations. Some of the damage is repaired, some is not. Lethal doses cause interference with transcription of some enzymes. In liver none of these effects of radiation is associated with the release of acid hydrolases. In spleen of irradiated animals, interference with transcription and DNA and even the appearance of pyknosis or karyorrhexis occur up to 30 minutes before the release of acid hydrolases.

The administration of puromycin does not release or change the concentration of acid hydrolases, and lysosomes appear hours after the block of protein synthesis. The administration of carbon tetrachloride leads to an interference with protein synthesis, which in turn results in triglyceride accumulation. These changes occur before the appearance of secondary lysosomes and the release of acid hydrolases.

One could extend the list to include circumstances in which the appearance of hydrolases follows other types of damage, but such an endeavor would be futile.

If in addition to the facts just described, we consider the difficulties of interpreting data on lysosomal enzyme release in necrosis or physiological remodeling of tissues because of the inescapable participation of polymorphonuclears and macrophages, which are rich in lysosomal enzymes, it seems fair to conclude that if lysosomal enzymes are likely to be involved in all forms of cellular death, they do not trigger it but scavenge the remains long after the lesions have become irreversible.

Correlation of Some Morphological and Biochemical Events in Cellular Death

On the basis of previous work by other electron microscopists, a group of Chicago investigators proposed an ingenious generalization referred to as "focal cytoplasmic degradation." These investigators postulated that the cell segregates pieces of cytoplasm for autodigestion by surrounding them with membranes. The cytoplasmic components are progressively digested within the confines of the membrane, and dense bodies are formed as a result. The concept provided an alternative, which continues to be overlooked, to the traditional lysosomal concept. The concept also raised meaningful questions about the origin of the segregating membrane, the mechanism of conversion of a double into a single membrane, the origin of the hydrolases, and the mechanism of elimination of the dense bodies. Even more far-reaching questions concern the factors which determine that degradation will take place in a restricted area and the type of signal that tells the surrounding healthy cytoplasm that a portion of its being is to be digested.

The concept of focal cytoplasmic degradation does not exclude the possibility that lysosomal enzymes could be transferred directly from the endoplasmic reticulum membranes into the portion of cytoplasm to be digested.

Although we do not imply that such a mechanism prevails under all circumstances, it is likely to obtain when cells die in part or as a whole. The concept has the advantage of simplicity, and, if correct, it would suggest that the hydrolases associated with the endoplasmic reticulum have physiologic functions that remain to be discovered. Recently this hypothesis has received support from electron microscopic studies in *Tetrahymena*. The implication is that acid hydrolases are not set apart to execute cells but play important physiological roles, and such roles may directly involve the enzyme associated with the endoplasmic reticulum.

It is now clear that the lysosome, a child of the test tube which became a constellation of particles when seen by the electron

FIGURE 9-1. Flow of metabolites in the cytoplasm.

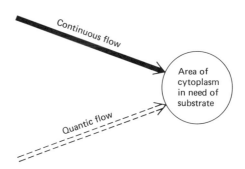

microscopist, participates in multiple physiological functions. But we must return to the mechanism of cellular death.

Can we in any way guess why some areas of cytoplasm become segregated? Let's assume that there are no qualitative but only quantitative differences in the mechanism of formation of focal cytoplasmic degradation in cells that survive the process and cells that die. Is it also not fair to assume that the concentration throughout the cytoplasm of macromolecular or other components needed in the cytoplasm for maintenance of the integrity of the membranes, replacement of structural protein, or other essential events is not uniform and that the supply of such components in an area in need may be not continuous but quantic (Fig. 9-1)? Therefore,

FIGURE 9-2. Hypothetical mechanism of cytoplasmic degradation.

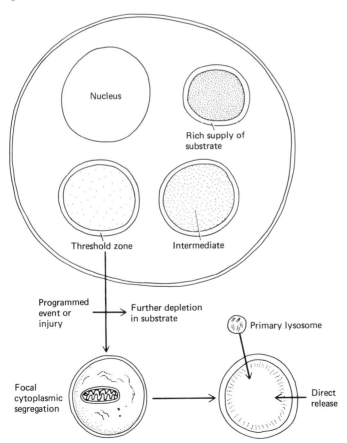

although some areas may be freshly supplied with the needed substrate, others may be close to depletion. These are threshold zones. Other parts may be somewhere in between abundance and famine. Many times because of a programmed event or an injury, the supply of substrates to the threshold zones is reduced, switching the equilibrium from anabolism to catabolism. Such areas are then segregated from the still healthy cytoplasm by membranes from which hydrolases are released (Fig. 9-2).

CONCLUSION

A great number of macromolecular lesions can cause cellular death. The steps that lead from the primary injury to death are not known. However, when a cell dies, in part or as a whole, areas of cytoplasm are segregated for autolysis, and at some point in their development the segregated areas contain hydrolases. These hydrolases may be introduced by fusion with primary lysosomes or by direct release of the enzyme from the endoplasmic reticulum into the autolytic vacuole. When hydrolases appear, the equilibrium in the segregated area is switched from anabolic to catabolic pathways, and the trend toward autodigestion becomes irreversible.

Numerous macromolecular alterations always precede autodigestion. Therefore, it would seem that they either trigger cell death or are located in the sequence of events that follow application of the injury somewhere between the primary alteration and the scavenging of the dying cytoplasm.

In any case, if we wish to understand the mechanism of cellular death, we must focus on those events that precede the development of focal cytoplasmic degradation, identify the first manifestation of injury, determine what injuries are critical in causing death, and discover the extent to which those injuries are reversible. Only such an approach can help the physician reach his ultimate goal: to prevent and cure disease. This goal is well expressed in the verse of Shakespeare, "find her disease/ And purge it to a sound pristine health" (*Macbeth* act 5, scene 3, line 52).

10

THE GREAT KILLERS, ATHEROSCLEROSIS AND CANCER

INTRODUCTION

The Egyptians embalmed millions of bodies, but they left us no account of what they found. Archeologists thousands of years later established that atherosclerosis and cancer existed then as they exist now. Up to the 11th century atherosclerosis and cancer must have been curiosities found at autopsy rather than major causes of death. Most people then died of pneumonia, tuberculosis, and epidemics. Today millions of people no longer die at one time, but in the technically developed countries humans die one by one as a result of heart disease, stroke, or cancer.

As was discussed in Chapter 1, sound pathogenic mechanisms were proposed only in the latter portion of the 19th century. The development of the cellular theory of disease gave us new insight into the mechanisms of cellular death, of reaction to infectious agents, and of cancer (although Virchow claimed that all cancers were of mesenchymal origin). The discovery of bacteriology and immunology identified the causal agent of many diseases and in many cases permitted effective prevention through vaccination. In the first half of the 20th century most vitamins and hormones were identified, and during World War II antibiotics were discovered.

As a result of this unprecedented acquisition of knowledge about disease mechanisms, the great epidemics were wiped out and diseases caused by defective nutrition were eliminated at least in the most prosperous parts of the world.

After 1950 the introduction of biochemical techniques in the study of disease clarified the pathogenesis of many hereditary and some endocrine diseases, but such an approach has seldom yielded effective cures. Nevertheless, amniocentesis has helped to prevent the occurrence of some hereditary diseases, and hormone replacement has permitted treatment of endocrine diseases.

Clearly, much remains to be done to understand the pathogenesis of numerous degenerative diseases of the nervous system, of common ailments such as peptic ulcers, of the ubiquitous hypertension, and of many bone and joint diseases. In fact, the list of diseases whose pathogenesis remains to be clarified is endless. In modern societies the two major killers are atheroscelrosis and cancer, and therefore these diseases deserve special consideration.

ATHEROSCLEROSIS

We shall not use the term arteriosclerosis (which is sometimes considered to be synonymous with atherosclerosis) because in the last decades there has been a tendency to use arteriosclerosis as a generic term covering a number of different diseases of the arteries. Atherosclerosis specifically refers to lipid and calcium deposition in the arteries and is associated with fibrosis and necrosis. The process ultimately causes focal hardening and narrowing of the arteries and formation of thrombi at the site of the lesion.

Pathogenesis

Atherosclerosis affects many arteries; it is more frequent in the abdominal than in the thoracic aorta. In the aorta or other large vessels it may cause weakening of the wall and the development of aneurysms, but the major cause of death as a result of atherosclerosis is progressive narrowing of one or more coronary arteries (the arteries that vascularize the heart) or of the arteries that bring the blood to the brain. Arterial narrowing (Fig. 10-1) leads to ischemia of these vital organs and results in myocardial or brain infarcts that severely incapacitate or kill the victim.

In spite of a massive research effort in this country and abroad covering the epidemiology, blood chemistry, and morphology of atherosclerosis, we have little knowledge of its pathogenesis. We do not even know for sure what sequence of morphological events leads to the formation of atheromatous plaques. Certainly yellow fatty streaks appear at an early stage. It is likely that some streaks expand whereas others dissapear. Lipids are believed to stimulate the proliferation of smooth muscle cells, which in turn are responsible for fibrosis. Some believe that only clones of smooth muscle cells are stimulated to proliferate, as happens in the development of a benign tumor. Sooner or later the lipid-laden cells die, the surface of the artery becomes ulcerated, calcium is deposited giving the artery an eggshell consistency, and thrombi form on the ulcerated surface.

Coronary Heart Disease

The left coronary arteries and their branches are occluded about twice as frequently as are the right coronary arteries. In both arteries, the first third of the artery after the origin is most often occluded. The major blood vessels of the heart are shown in Figure 10-2.

The consequences of coronary artery occlusion are varied; they result mainly from the speed at which occlusion occurs and the existence of an anastomosis at the time of occlusion. If the occlusion is progressive, the two coronaries may anastomose and no severe consequences will ensue, except that eventually some repeated episodes of ischemia may develop when oxygen requirements are increased, leading to external pain called angina pectoris.

Occasionally, a narrowed artery may be occluded completely by a thrombus, embolus, hemorrhage, or rupture of an atheromatous plaque. The artery may be occluded so suddenly that the patient

FIGURE 10-1. Progressive occlusion of artery by atherosclerosis.

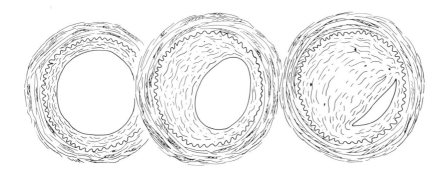

dies within 12 hours, before anatomical injuries are observed in the myocardium.

Coronary obstruction most often evolves into myocardial infarct; the site and extensiveness of the necrotic process depend on the anatomic distribution of the occluded artery. When the descending branch of the left coronary artery is occluded, infarcts occur in the anterior and lateral walls of the left ventricle. Right coronary occlusion leads to infarcts of the posterior third of the septum and the posterior part of the left ventricle. Occlusion of the left or right circumflex results in infarcts of the lateral wall of the left ventricle. Depending on the height at which the occlusion occurs in the arterial lumen, infarcts of the atrium or of the right ventricle may complicate left ventricular infarcts.

The classic pathological appearance of the myocardial infarct can be seen in victims of coronary occlusion who died 2–15 days after the infarct. Although the disease is called myocardial infarct, it usually involves the three layers of the heart. At gross examination, a fraction of the heart is a tawny yellow-brown color, and it is, more than normal myocardium, friable on sectioning. The surface section is homogenous, and individual bundles of fibers can no

FIGURE 10-2. Vasculature of the heart.

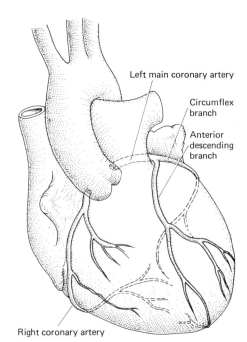

longer be recognized. The pericardium is usually covered with a thin fibrinous layer, and the endothelium with laminated thrombus. Histologically, three distinguishing features can be observed: (1) muscular degeneration characterized by hyalin degeneration of the fibers and pyknosis of the nuclei; (2) patches of hemorrhage throughout the infarcted area; and (3) polymorphonuclear invasion, especially at the edges of the infarcted area. The nucleus of the fiber may increase in size and become polyploid, or the fiber may become multinucleated. The necrotic area is usually populated with hemosiderin-loaded macrophages. Six to eight months after the myocardial infarct, the necrotic area is entirely replaced by fibrotic tissue.

The patient often dies from forward failure soon after the infarction, but sometimes he survives, and then his life is threatened by the presence of mural thrombi (these do not organize; thus they are subject to autolysis, during which they constitute a permanent danger of emboli) or by rupture of the heart. The chances of rupture depend on the extensiveness and stage of development of the infarct. The more softened the infarct, the greater the chance of rupture. Hemorrhage and polymorphonuclear infiltration are responsible for intramural softening of the heart. Naturally, hemodynamics are a determining factor in heart rupture. A person with a blood pressure over 140 mm Hg is three times more susceptible to heart rupture than a normotensive person.

If the infarct is not too extensive and the patient can be maintained under complete rest to avoid stress on the heart muscles, healing may occur. Healing does not consist of regeneration of the muscle fibers to restore the structural integrity of the heart, but it results in hypertrophy of the fibrous tissues associated with hypertrophy of some of the remnant fibers. In any case, these healed areas constitute weak points in the heart wall that bulge when blood pressure is high, leading to the formation of sac-like areas called aneurysms of the heart. The aneurysms are favorite sites for mural thrombi formation. Of all cases of sudden death, 65% are due to cardiac failure, and 65% of the latter are due to coronary artery disease.

The clinical observations in acute myocardial infarct are revealing. An elderly individual, usually a man, experiences a sudden chest pain that cannot be relieved by the administration of vasodilating drugs. Severe dyspnea often develops. Ischemia impairs the heart's function, and blood pressure drops considerably, even in individuals with high blood pressure. The heart sounds are feeble, often with galloping rhythm and other arrhythmias. Furthermore, the condition for electrical conduction in the heart is modified, and the electrocardiographic pattern is altered.

The electrocardiographic findings are not always typical, but usually they develop within 2–12 hours after the onset of ischemia. The details of the electrocardiographic changes cannot be described here. Suffice it to point out that they vary depending on the location of the infarct in the heart walls. The extensive necrotic process causes some characteristic changes in the blood: polymorphonuclear accumulation, a drop in the sedimentation rate, and an increase in serum transaminase levels. (Coronary atherosclerosis and previous occlusions were found in 90% of the hearts in a series of patients with angina pectoris who died of noncardiac causes.)

Associated with the increased levels of blood transaminase is an increase in catecholamines, particularly norepinephrine. But norepinephrine levels may be increased in cases of angina pectoris, whereas transaminase is not increased under such conditions.

Cerebral Consequences of Atherosclerosis

Stroke is the main consequence of cerebral arteriosclerosis, and it may be caused by hemorrhage or obstruction of an artery. Stroke by hemorrhage usually occurs in patients who have severe hypertension, and it may or may not be associated with arteriosclerotic lesions of the vessels. In any case, the primordial mechanism of injury in stroke by hemorrhage is hemodynamic stress on the artery rather than any local condition.

Brain infarcts result from progressive or abrupt occlusion of the cerebral arteries. The pathogenesis of this injury is best understood if the vascularization of the brain is reviewed. The vessels of the brain stem from three main sources: the two carotid arteries and the vertebral arteries. The right and left carotid arteries divide at the level of the external angle of the optic chiasma to yield the anterior and median cerebral arteries. The vertebral arteries fuse to yield the basilar trunk, which divides to yield the two posterior cerebral arteries. The anterior, median, and posterior cerebral arteries are welded in a continuous circle, called the polygon of Willis, by the anterior communicating artery connecting the two anterior cerebral arteries, and by the posterior communicating arteries connecting the posterior with the median cerebral arteries.

Obviously, the location of the brain infarct depends on the site of the occlusion (the vasculature of the base of the brain is shown in Fig. 10-3).

Embolization is responsible for sudden occlusion, but thrombosis usually causes the slower forms of occlusion. The middle cerebral artery is one of the arteries most frequently affected by thrombosis. Thrombosis and emboli formation lead to the development of an infarct, the size of which depends on the location of the injury in the arterial tree and on the presence of an anastomosis. Autopsy findings show that the arteries of the circle of Willis are favorite sites for arteriosclerosis. The plaques may lead to progressive narrowing of the vessel, with localized anoxemia or thrombosis. In contrast to emboli, thrombi affect the right and left cerebral hemispheres equally. The occlusion produced by the thrombus is usually more progressive than that resulting from emboli, although it is not necessarily less extensive. However, sometimes the occlusion caused by thrombi may occur rather suddenly, and this usually results from associated spasms.

The superimposed spasm also explains how full or partial recovery is sometimes possible after thrombic occlusion. Thus, the focal symptoms are usually preceded by a period of dizziness, headache, lack of concentration, and uncoordination. The symptomatology depends on the nature of the artery that is obliterated: if the anterior cerebral artery is affected, no significant symptomatology follows; if the posterior cerebral is affected, blindness de-

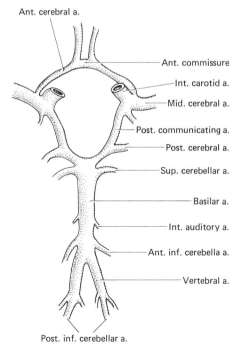

FIGURE 10-3. Vasculature of the base of the brain.

velops; if the median artery is affected, hemiplegia and hemianesthesia develop. If collateral branches are affected, various motor or functional systems are damaged.

Brain Infarct The site of a brain infarct depends on the artery that is obstructed, and the extensiveness of the injury is a function of the density of the interarteriolar anastomoses and their degree of permeability. If the anastomoses are narrowed by arteriosclerotic processes or other types of injuries, the infarct will be extensive. Superficial infarcts are cone shaped with the base at the surface and the apex at the center; central infarcts are spherical.

Pathologists describe brain softening in three stages: white, red, and yellow. Such a subdivision is somewhat artificial because it is difficult to delineate one stage from another; however, such a classification has didactic advantages.

The period of white softening is short; during that time the brain bulges somewhat at the surface and an area of softening, although difficult to recognize on fresh material, becomes obvious after fixation.

Red softening results from a complex interaction of necrosis, hemorrhage, and lymphocytic infiltration. During red softening the neural cells degenerate, the myelin sheaths are fragmented, and their lipid components are phagocytized by invading macrophages and polymorphonuclear leukocytes. The entire brain substance autolyzes. This leads to hemorrhage and bulging of the involved area at the surface. The convolutions are flattened, and on sectioning the area appears brownish and friable.

In yellow softening, the entire mass of the brain is softened and replaced by a brownish-yellow material stained by the blood pigment. The infarcted area is not repaired, so at the end of the process when all tissues are autolyzed, all that is left is a shrunken cystic area surrounded by glial proliferation. The brain is smaller than normal, and the convolutions, particularly those in the frontal and parietal areas, are decreased in size; this is indicated by widening of the sulci.

Encephalomalacia When thrombi, emboli, or other forms of vascular occlusion affect small arteries, they cause encephalomalacia. This results from occlusion of smaller vessels irrigating the cerebral cortex, the striated nuclei, or the internal capsule. Occlusion of these terminal arteries leads to degeneration of the nerve cell, proliferation of glial cells, and fragmentation of the myelin sheath, the lipid components of which are phagocytized by macrophages. Under some circumstances, such reactions may lead to the development of a small glial scar or a cyst surrounded by a glial capsule.

Particularly susceptible to encephalomalacia are the striatum, the hippocampus, the dentate nucleus, and the cerebral cortex.

Obviously, encephalomalacia progresses slowly. For years no symptomatology is detected, but with time the patient loses memory, especially recent memories, his moods change, and he becomes tired rapidly and shows marked nervous depression. If the injuries are mainly in the corpus striatum, pseudobulbar paralysis may develop. The classic patient is an older person with an immobile face and a wide-open mouth (saliva drips from the corners of his lips), and he has difficulty swallowing food and articulating. He is excessively sensitive and cries or laughs easily. Lesions that cause exaggerated tendon reflexes and a positive Babinski reflex are often associated with these bulbar signs.

Cerebral Hemorrhage The exact mechanism of the development of cerebral hemorrhage (Fig. 10-4) is not so well established as to lead to unanimous agreement on its pathogenesis. At least three theories have been proposed. The first assumes that cerebral hemorrhages are of nervous origin. It postulates that temporary excitation leads to vasodilation followed by vasoconstriction, during which an unknown mechanism causes red cells to travel from the vascular lumen into the extravascular spaces. The amount of red cell extravasation depends on the duration of the excitation. The second theory proposes that the mechanism triggering the cerebral hemorrhage resides in a vascular spasm, which leads to anoxemia followed by necrosis of the vascular walls and consequent hemorrhage. The third theory contends that the hemorrhage originates from the rupture of the vessel, either in areas weakened by arteriosclerosis or by the presence of a congenital miliary aneurysm. Sudden ictus

FIGURE 10-4. Cerebral hemorrhage (*shaded area*) in left hemisphere.

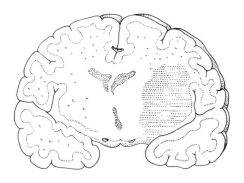

may well result from such a mechanism. Indeed, in ictus the hemorrhagic area does not truly infiltrate the nerve tissues and does not result from the confluence of smaller hemorrhages, as suggested by the previous hypotheses. The focus of hemorrhage in ictus is homogenous and pushes the normal nerve tissue toward the periphery.

Obviously, hemodynamic factors also determine the site and the time of the rupture. Consequently, cerebral hemorrhages occur more frequently among patients with hypertension, and at times of excitement or episodes of coughing, sneezing, vomiting, or sometimes after a hearty meal or a cold bath.

Again, the branches of the median cerebral artery are most often the site of cerebral hemorrhage. The hemorrhage usually leads to apoplectic ictus, the patient falls, and severe coma follows. When he awakens, the patient is usually hemiplegic. His face is red, he is sweating, and he is breathing noisily or snoring. His pupils do not react to light and are dilated. Conjunctival and swallowing reflexes are missing. If he recovers, his reflexes reappear and his temperature increases, and when the coma has disappeared, the patient shows focal signs resulting from destruction of the basal ganglia and the internal capsule with total anesthesia and hemiplegia. It would be most useful to be able, on the basis of the evolution of the hypersensitive syndrome, to predict the chances of death from cerebrovascular accidents, because under those circumstances it might be possible to prevent the occurrence of the accidents. Unfortunately, this is impossible. In studies of large numbers of hypertensives who had cerebrovascular accidents, a preclinical history gave no clue as to the chances of developing stroke.

Concluding Remarks on Atherosclerosis

Despite the difficulties encountered in identifying the epidemiological parameters of atherosclerosis, the clinicopathological correlations between coronary heart disease detected by electrocardiography and blood cholesterol have established two important facts: (1) environment affects the incidence of atherosclerosis in a number of different ways—diet and stress play key roles in influencing the risk of developing coronary heart disease; (2) although we do not know what level of blood cholesterol is safe, there is a close correlation between blood cholesterol levels and the incidence of atherosclerosis.

Epidemiological studies have made it possible to describe the profile of high-risk individuals. Coronary heart disease is most likely to affect a man of average height who is slightly obese, middle-aged, and has high blood cholesterol and high blood pressure. Such individuals usually eat too many rich foods, smoke, and are physically inactive. Because they are ambitious, they have many responsibilities, are anxious about them, and are constantly faced with new deadlines.

Using blood cholesterol as a measure of risk, it is possible to manipulate the environment in the hope that such manipulation will reduce the incidence of atherosclerosis. Consequently, any time a person is close to the atherosclerotic profile described above, efforts should be made to modify his environment so that his profile differs from that of the typical atherosclerotic. The candidate for heart disease should eat less, learn to relax, engage in controlled exercise, give up smoking, and plan ahead.

Because of the complexity of the pathogenesis of arteriosclerosis, it seems futile to attempt to weave into a coherent scheme all the factors involved in the pathogenesis of the disease.

CANCER

Clinical Pathology of Cancer

In Chapter 5 we discussed one of the potential pathogenic mechanisms of cancer. Cancer has become such a major cause of death that it deserves further study. We shall consider what it looks like grossly and what may cause it.

Atossa, the daughter of Darius, is believed to have been cured of breast cancer by cauterization. What happened to Atossa to cause cancer and what would have been the course of her disease if she had not been treated? Nothing is known of the agent that causes cancer of the breast in humans. Yet we can describe the evolution of the process.

The breast is a glandular structure whose major role is to secrete milk. Ducts that open in the nipple ramify into smaller and smaller ducts that ultimately open in the gland that secretes the milk. These are round structures lined by secretory cells, resting on a membrane made of collagen and polysaccharides, the basement membrane. When a cancer of the breast develops, one or more cells of either the duct or the acini are converted from normal into cancer cells by a virus, chemical, or some other agent still not identified.

The cancer cell is a rebellious cell, it does not respond to intracellular or extracellular mechanisms of regulation. The balance between cellular growth, maturation, and cell death is disturbed and the cell proliferates without check. Normally in the breast, as in other tissues, there are well-delineated borders between the epithelial (the secretory and the ductal cells) and connective tissue. In cancer these borders are no longer respected. Some cells of either the ducts or the milk glands escape normal controls. These cancer cells proliferate and invade the connective tissue not only by pushing it out of its way, but by actually digesting, infiltrating, and replacing it.

The physical appearance of the breast changes; the nipple is pulled within the mass of the breast, and the surface of the skin forms tiny dimples that give the skin the appearance of orange skin. The physician and sometimes the patient may be able to palpate a mass in the breast. If this were the extent of the disease, the consequences for the victim would be limited. Excision of the tumor would cure the patient. Unfortunately, the disease does not stop locally.

The cells migrate by invasion of the vasculature, in particular the lymphatics and the blood vessels. This migration is called metastasis. The metastasis appears first in the lymph nodes of the axilla and the subclavicular regions. With time the cells penetrate the blood vessels and lodge in the bones and in the lungs. The cells thrive on the building blocks of the host, trapping glucose, nitrogen, and other nutrients. The patient dies as a result of invasion of vital organs by metastasis and exhaustion of the body's resources in a process called cachexia.

Let's consider another example, cancer of the lung. Here we know one cause—smoking cigarettes. However, it is not so clear what substances in cigarette tar are responsible for cancer and whether or not other environmental factors contribute to the development of cancer. Lung cancer used to be prevalent in men and rare in women, but its incidence in women has been increasing as more women have started to smoke. The cancer starts with changes in one or more cells in the epithelium of the bronchi. The bronchial cells—like those of the breast in breast cancer—escape normal regulation, invade the walls of the bronchi, spread to lymph nodes and blood vessels, and ultimately kill the patient.

We can now return to our original definition of a cancer cell. It is a cell that escapes some internal and external mechanisms of control of cellular proliferation and maturation, and in the process it acquires survival advantages over most cells of the host.

We do not know whether only a single cell that later multiplies is converted to a cancer cell, or whether several cells are changed at once and each is the parent of a new cancer focus.

Pathogenesis of Cancer

Neither do we know how often a normal cell is converted into a cancer cell but is unsuccessful because of host defenses, cellular immunology being principal among them. Histologic observations in humans and animals have given us information on how many cells are involved in malignant transformation at one time. In animals carcinogens that produce liver cancer induce numerous nodules, suggesting that several cells have undergone the conversion process from normal to cancer cells. In early bladder cancer in humans one often finds multiple foci of intraepidermal cancer cells (cancer *in situ*) suggesting again that more than one cell was converted to the malignant state.

It is more difficult to determine if host defenses are mounted against malignant transformation. On the basis of circumstantial evidence not always convincing, Thomas has proposed that cellular immunity was evolved to survey the organism for the appearance of cancer cells. The process is referred to as immunosurveillance. Thus, a special population of lymphocytes, T lymphocytes, patrols the body, identifies the cancer cells, and kills them.

What is certain is that the properties of the cancer cell are transferred from one generation of cells to another. Translated in molecular terms, this means that in the process of conversion of a normal into a cancer cell, the cancer cell undergoes a change in its pattern of gene expression which provides it with the survival advantages described above. Since the changes are transferred from one generation to another, one may expect that they are permanently imprinted in the genome. The genome is primarily composed of nucleic acids and proteins, and at a first approximation it would seem quite logical to assume that cancer is caused by a direct alteration of either DNA, proteins (histone or nonhistone proteins), or the DNA-protein complex. Surely these possibilities are not excluded. But it is also possible that the initiating event in cancer involves other macromolecules such as those that compose the cell membrane. A feedback loop between the membrane and the genome could ultimately be responsible for permanent alteration of the genome.

The term "initiation" has a special meaning in carcinogenesis. There is evidence that cancer develops in two stages, initiation and promotion.

Extensive studies of two-stage carcinogenesis were undertaken in Berenblum's laboratory. The classic experiment consisted of a single application of methylcholanthrene followed by repeated applications of croton oil. Such treatment yielded skin cancer. If the croton oil was applied before methylcholanthrene, no cancer developed. Croton oil alone is noncarcinogenic. A single dose of methylcholanthrene produces only a few tumors, but subsequent administration of croton oil increases the incidence of cancer and reduces the latent period for the appearance of cancer.

On the basis of these experiments, it was concluded that methylcholanthrene acts as an initiator, whereas croton oil acts as a promotor. It seemed logical to assume that initiation induced a cellular injury that remained dormant until cellular proliferation took place. In such a case, one can anticipate that the incidence of tumors would depend on the amount of promoter applied. Other examples of two-stage carcinogenesis have been discovered. The systemic administration of urethane (initiator) followed by the topical application of croton oil (promoter) leads to the development of skin cancer at the application site.

Two-stage carcinogenesis is not restricted to the skin. In some strains of rats, cancers develop in ovaries implanted in the spleen only if a carcinogen is also administered. In some mice, leukemias can be produced with a single dose of X-irradiation (initiator) followed by the administration of urethane (promoter).

The dissociation of the mechanism of carcinogenesis into two distinct steps has permitted investigators to determine whether the factors that modulate carcinogenesis act on the latent period or on promotion. For example, the reduced incidence of tumors because of low caloric intake or the increased incidence of tumors caused by hormone administration appear to result from an influence on the promoter stage. The active promoting factor in croton oil was isolated. The factor (12-O-tetradecanoylphorbol-13-acetate) is a fatty acid ester of a polyfunctional diterpin alcohol.

We know little of the molecular events associated with initiation or promotion. The general view is that initiation corresponds to a permanent molecular alteration of the cell, whereas promotion results in cell proliferation. Little is known of the permanent change that occurs during initiation, but it cannot be excluded that binding of the carcinogen to DNA or other macromolecules may be responsible for the changes.

Chemical Transformation in Vitro

Various carcinogens have been used to achieve transformation *in vitro*, including polycyclic hydrocarbons, alkylating agents, and others. Transformation is accompanied by a number of alterations of the morphological, biochemical, and functional properties of the cell. These changes include alterations in: membrane transport, membrane structure, adhesiveness to other cells and substrata, chromosomal number, growth characteristics, serum requirements, and morphological features. The relationship between any of these alterations and the transforming event is not known.

Carcinogens in Humans

Certainly studies of experimental cancer have taught us a great deal about the biochemistry of cancer and the molecular changes that convert innocent chemicals into carcinogens (see Chap. 5). We will not dwell further on cancer in animals, but instead we will discuss causes of cancer in humans. Undoubtedly, the combined efforts of organic chemists, biologists, and biochemists have dug deep tunnels into the unknown of cancer. Chemicals discretely but fatally pollute the water we drink, the food we eat, the air we breathe, and the smoke we inhale. Once such chemicals have been introduced into the body by mechanisms unknown and difficult to unravel—in part because of the multiplicity of factors that may affect carcinogenicity (hormones, age, other carcinogens, etc.)—these chemicals may convert one or more intact cells into cancer cells.

To deny that some of the carcinogens proved to be efficient in the laboratory are active in humans is unrealistic and dangerous. To be convinced, one needs only to review epidemiology and the geographic distribution of cancer. In 1853 Paget vigorously attacked Sir Percivall Pott's interpretation of the pathogenesis of cancer of the scrotum in chimney sweepers and claimed that soot had nothing to do with cancer. Years later, Curling described a higher incidence of skin cancer among gardeners who used soot as a fertilizer. Workers who used tar to build ships, roads, and shoes frequently had warts that over the years developed into cancer. The warts sometimes appeared in unusual parts of the body, such as the ears of those carrying on their heads sacks full of coal, or the feet of those working barefoot. The weavers of Lankashire, whose clothes were continuously splashed with lubricating oil, developed skin cancer at the site of maximum impregnation of their clothes. Paraffin workers developed carcinomas of the scrotum, forearm, and leg.

Perhaps the most dramatic illustration of the presence of carcinogens in the environment was observed among workers inhaling aniline vapors. In 1895 Rehn described several cases of cancer of the bladder among such workers. But before that, Granhomme had listed hematuria among the symptoms resulting from chronic aniline intoxication. This was less than 20 years after aniline dyes were first used in industry. Soon after Rehn's description, cancer of the bladder was observed among workers in many factories in Germany and Switzerland. Kennaway and his associates in 1921 reported

5614 cases of cancer of the bladder and demonstrated that the incidence was four times greater among aniline workers than among farmers. The disease was practically unknown in the United States until the turn of the century, which may well reflect the fact that the aniline industry was introduced on the American continent only in 1893. That the repeated application of crude material was related to the development of cancer could no longer be doubted once it became possible to produce cancer experimentally with the aid of the crude material and to extract pure carcinogens from most of the materials described above.

This brief outline by no means exhausts the large list of profession-related cancers due to chemical carcinogens, but it illustrates one of the most threatening socioeconomic problems of our times. The list is endless. Those working in coal mines frequently develop cancer of the nostrils. Asbestos workers develop mesotheliomas of the pleura. Workers exposed to benzol may develop leukemia. Those who work in arsenic smelters develop skin cancer.

In 1971 the development of cancer in animals exposed to vinyl chloride was reported. Vinyl chloride is a colorless gas, barely soluble in water, but slightly soluble in organic solvents such as ethanol. Polyvinyls contain 2 to 7% of vinyl chloride.

Talc, a magnesium silicate, has been implicated in the high incidence of gastric carcinoma in Japan. Indeed, the Japanese consume a great deal of rice that is coated with glucose and talc to produce a nutrient superior in appearance and taste. Talc is frequently contaminated with several types of fibrous silicates usually identified as asbestos.

The potential relationship between the development of benign hepatic adenoma and the use of contraceptives has been suggested.

When the relationship between the agent and cancer is obvious, as in the case of aniline and cancer of the bladder, prevention is easy—especially if the responsibility is social rather than individual. Sometimes even when the relationship between agent and cancer becomes convincing, all efforts to protect individuals fail. A case in point is the relationship between smoking and lung cancer. This type of cancer occurs almost exclusively among heavy cigarette smokers. Those who smoke pipes and cigars and do not inhale the smoke suffer few effects. Today more people, and especially more young people, die of lung cancer in the civilized world than in all recorded medical history.

Because there is a great similarity between modes of consumption of marijuana and tobacco, it is important to determine whether or not the smoking of cannabis is carcinogenic. Compared to cigarette smokers, even heavy users of marijuana inhale a small amount of smoke. Therefore, unless marijuana contains high concentrations or an unusually effective carcinogen, it is unlikely that the incidence of cancer in humans caused by marijuana will be very high. Nevertheless, the consequences of our lack of knowledge of the dangers of smoking cigarettes have been so tragic to the human race that it seems imperative to ascertain the carcinogenic potential of any drug likely to be widely used.

Our environment probably contains many unsuspected carcinogens. For example, aminoazotoluol, extracted from seemingly harmless buttercups, was used for coloring margarine, and was proven to be carcinogenic. Cosmetics, foods, and drinks are colored artificially; which coloring materials are carcinogenic? The carcinogenic properties of acetylaminofluorene were discovered by testing insecticides that were sprayed on cranberry bushes until the 1960s. Cyclamates are believed to be carcinogenic for the bladder. Carcinogenic polycyclic hydrocarbons have been detected in the smoke emanating from charcoal grills and have been found on the surface of broiled meats. The number of carcinogens in our environment is unknown, yet detecting them may be more beneficial to man's health than curing cancer, because once the cause is known, preventive measures can be taken.

The carcinogenic properties of nitrosamines are well established. Both nitrites and secondary amines can be found in the diet. Gastric juice can stimulate the formation of nitrosamine by the combination of nitrites and the secondary amines; similarly, enteric bacteria can form nitrosamine from nitrites and the secondary amines. At least one nitrosamine, dibutyl nitrosamine, is known to be carcinogenic for rodent bladder.

Whether a similar mechanism plays a role in carcinogenesis in the human gastrointestinal tract remains to be seen. Although the incidence of stomach cancer is reaching epidemic proportions in Japan, it is decreasing in the United States.

The geographic distribution of cancer of the liver is somewhat unusual. Practically nonexistent in some countries such as the United States, it is common in other parts of the world, e.g., Africa. In Mozambique the incidence of cancer of the liver is said to be 500 times that in the United States and 30 times that in Johannesburg. In Hong Kong, cancer of the liver is found in 30% of all autopsies. Carcinogens produced by fungi (aflatoxin) could be responsible for the high incidence of cancer of the liver in those areas. In many of the countries where cancer of the liver prevails, aflatoxins have been found to contaminate the food. In addition, a number of other fungal toxins, often of quite different molecular structures, have been found to cause cancer not only of the liver, but also of the kidney and the stomach in experimental animals.

The list of drugs that are either proven or potential carcinogens increases every day. There is a serious question as to the propriety of their use in humans. An impressive number of antiseptics, antibiotics, analgesics, tranquilizers, and drug additives are known to be carcinogenic in animals or in humans when humans are exposed to these compounds because of their professional activities.

Tars have been part of the therapeutic arsenal of dermatologists for generations. Chlorinated hydrocarbons, chloroform, and tetrachloroethylene have been used for many years as anesthetics; tetrachloroethylene has been used as an anthelmintic. Cancer of the liver was reported in an individual using carbon tetrachloride as an extinguishing fluid. Whether chlorinated hydrocarbons are frequently involved in human carcinogenesis remains to be established.

Diethylene glycol, frequently used as a solvent for drugs, produces cancer of the bladder in rats. The carcinogen β propiolactone is used as an antiseptic to sterilize arterial grafts and serum for transfusion. Propionolactone produces fibrosarcoma when injected, carcinoma of the skin when painted, and carcinoma of the lung when introduced intratracheally in rats.

β-Propionolactone, which contains a four-membered, strained ring, is highly reactive. It is an alkylating carcinogen. For example, it reacts with the sulfhydryl group of cysteine to yield S-2-carboxyethyl cysteine.

Several petroleum products, such as mineral oil and petroleum waxes, may or may not contain carcinogenic hydrocarbons, depending on their degree of refinement. A number of dyes (trypan blue, acid green, acridine, and creosote) used as antiseptics for local application or in gauze preparations for treating wounds are carcinogenic in animals. Acridine derivatives, such as acriflavine and quinoline, have been suspected to be carcinogenic, but no conclusive evidence has been provided.

Phenol acts as a promoter in skin cancer produced in mice with dimethylbenzanthracene (DMBA). Distillation of coal tar between 200° and 400° yields a population of compounds referred to as creosote. Creosol is a major component of creosote. Creosote produces skin cancer in mice, and reports have appeared of skin cancer with metastases developing in people working in the creosote industry. Tannins, long used in treating burns, induce hepatomas in rats. Reserpine, used in the treatment of hypertension, markedly accelerates the induction of hepatomas in animals.

In 400 B.C., Hippocrates recommended arsenical ointments for the treatment of skin ulcers. The use of arsenic was advocated in an incredible number of unrelated diseases: anemia, bronchial asthma, eczema, acne, epilepsy, and psychoneurosis. The efficacy of using organic arsenicals in therapy for syphilis and some parasitic diseases such as amebiasis and trypanosomiasis has been known for a long time. Arsenicals have been added to food to prevent parasitic disease in fish, poultry, and livestock. Inorganic arsenic, which has been generously prescribed in ointments and potions for centuries, is carcinogenic in man. Indeed, arsenical dust and fumes cause cancer of the skin and lungs in the smelter workers. The drinking water of Liechtenstein is polluted with arsenic and it is believed to have affected the increased incidence of cancer of the skin and viscera in inhabitants of the small country. Increased incidence of skin cancer has been reported in patients with cirrhosis who are subjected to prolonged treatment with arsenic.

Both nickel and chrome, which are used in the preparation of nails and plates for bone surgery, are claimed by Heuper to be carcinogenic in man. Isoniazid, a drug that has helped eliminate tuberculosis, produces adenomas of the lung in mice. The intricacies of interpreting this finding will be reviewed later.

The discovery of the carcinogenicity of certain drugs raises the important and difficult problem of deciding whether a drug proved or suspected to be a carcinogen in animals or man should continue to be used in therapy. Moreover, should all drugs be tested for potential carcinogenicity, and, if so, what tests will be most revealing and will determine whether the drugs should be retained or withdrawn from the market?

First, we must remember that the concept of carcinogenicity is a relative one. A given carcinogen is not carcinogenic for all animals, all species, all tissues in an animal, and at all times during a life span. Thus, the carcinogenic properties of a chemical vary considerably with age, strain, species, and type of cancer. Consequently, if a test for carcinogenesis is positive in one strain of animal, it cannot be concluded that it will be positive for all strains. Moreover, if one method of administration of the carcinogen is used, it cannot be concluded that introducing the drug by other methods will also cause cancer.

In contrast, if the results of the studies in animals are negative, the possibility that the drug could be carcinogenic in humans cannot be excluded. Therefore, the potential carcinogenicity of a given drug should be tested in a number of different species (rats, mice, hamster, etc.), and testing should involve the entire life span of the animal (pregnant females, newborns, and adults). In all cases, the survey should include complete autopsies and examination of all lesions by a competent pathologist.

Two methods are available to determine if a drug is carcinogenic in humans: retrospective and prospective approaches. Both methods obviously resort to statistical analysis of large amounts of information. In the retrospective approach, the incidence of cancer among patients who received the drug for a long time is investigated. But drug dosage and circumstances of use are difficult to survey accurately (the difficulties involved are illustrated in the discussion of the carcinogenic role of therapeutic radiation).

In the prospective method, the survey is planned at the start, and follow-up studies of the patients and frequent examinations to detect cancer are planned. Such studies in human are complicated by: (1) the long latent period between the time of administration of the carcinogen and the development of cancer; (2) the multiplicity and diversity of cancers that occur; (3) the difficulty of obtaining a complete record of the intake of other drugs and of food additives—even if intake could be known, it is likely to vary considerably among individuals; (4) the complexity of carcinogenesis and its occurrence in two phases—initiation and promotion; and (5) the difficulty in determining whether a drug acts as a promoter or an initiator. Despite these difficulties, careful epidemiological studies have occasionally linked the development of specific cancers to a specific carcinogen (skin cancer and arsenic, lung cancer and cigarettes). In such cases, it is simple to propose the elimination of the carcinogen from the environment.

The case of isoniazid is of particular interest. As mentioned before, isoniazid produces adenomas of the lungs in Swiss mice; but when investigators attempted to demonstrate isoniazid's carcinogenicity in other species, they were unsuccessful. Moreover, careful examination of the animals receiving isoniazid revealed that as it induced lung cancer, the drug also reduced the incidence of some other cancer to which that strain of mice is prone (e.g., breast cancer). In most cases, however, the decision to maintain or eliminate a potential carcinogen from the pharmacopeia rests on complex and sometimes conflicting data. In the face of these many, sometimes irreconcilable, facts, how is it possible to enounce wise rules for the use of drugs? Obviously, no drug should be used unless it is indispensable to the maintenance of the patient's physical and mental health, and only when needed should drugs be used for securing temporary comfort (elimination of pain).

Special consideration should be given to all drugs that are used for prolonged periods of time, such as hypotensive agents and anticoagulants. If the drug is carcinogenic in animals, physicians should attempt to replace it with a noncarcinogenic drug whenever possible. Carefully planned studies should be pursued in patients submitted to prolonged therapy with new drugs, keeping the possibility of cancer in mind.

If a drug is a proven carcinogen in humans, it should be eliminated from the market unless, of course, the condition of administration excludes the possibility of carcinogenesis, or the expected life span of the patient (because of age or disease) is much shorter than the latent period necessary for developing cancer.

Drugs not known to be carcinogens in humans or rats should not be considered safe until prospective studies in humans have established them to be. The chemical structure of the drug may serve as a guide for the detection of potential carcinogens.

Special precautions should be taken when the drug is to be administered to pregnant or lactating women or to newborns.

Several observations suggest that human cancers might frequently be caused by a combination of factors. For example, it would appear that the consumption of alcohol by itself does not increase the incidence of cancer of the mouth or the esophagus, but the combination of heavy drinking and smoking increases the risk of cancer considerably.

Cancers and Viruses in Humans

Viruses cause cancer in amphibians, chickens, and mammals. In view of the widespread occurrence of viral carcinogenesis in animals, it seems unlikely that human cancers are never produced by viruses. However, there is no evidence that human tumors, benign or malignant, are caused by viruses (warts are an exception to this). Although conclusive evidence is still lacking, there is a growing suspicion that viruses may play a role in the etiology of human cancer. This assumption is based on: (1) the observation that cancer viruses occur in many animal species, including primates; (2) the possibility of producing cancer in animals or transforming cells in culture with nucleic acids or viruses obtained from human tumors; and (3) the epidemiology of some human cancers and the identification of viruses in many human cancers.

Oncogenic viruses affecting humans are expectedly difficult to identify for two reasons. First, in contrast to experimental animals, human populations are highly heterogeneous. Second, in humans, oncogenic or even nononcogenic viruses may work in combination with other carcinogens to produce cancer. The potential cooperative role that ordinary infectious viruses may play in human carcinogenesis by combining their effect with that of chemical carcinogens or X-irradiation has not yet been evaluated.

Perhaps the most compelling argument in favor of a viral etiology in some human cancers derives from epidemiological and electron microscopic observations. In fact, long before viruses were discovered, the Academie des Sciences of Lyon awarded its prize to Bernard Peyrilhe, a French surgeon who proposed, on the basis of some epidemiological observations, that cancer is infectious. The familial incidence and the geographic distribution of leukemia, Hodgkin's disease, and Burkitt's lymphoma are often invoked as evidence of an infectious etiology. Three observations support the hypothesis of the viral etiology of leukemia: the detection of clusters of cases, the coexistence of leukemia in children and their pets, and the high incidence of leukemia among children who have experienced repeated viral infections.

Clusters of cases have been reported in Texas and Illinois, and reports of coincidental hematological changes in children and their pets have led some to believe that domestic animals could serve as vectors for the virus. Viral particles have been found in sarcomas in cats, in leukemia in dogs, and in the milk of cows carrying lymphomas. Of course, if a viral etiology were established for leukemia, the relation of the virus to other pathogenic factors, such as X-irradiation, chromosome alteration, hormones, and chemicals, would need to be clarified.

The demonstration of reverse transcriptase activity in human leukemic cells has been offered as an argument for the viral origin of such cancers. However, the enzyme is found in normal lymphocytes as well.

More recently, Spiegelman provided more compelling evidence of a molecular nature for the viral origin of human leukemias by: (1) demonstrating the presence of a viral RNA homologous to that of Rauscher leukemia in leukemic cells; (2) establishing that the transcriptase is associated with a particle with the density of the Rauscher leukemic virus; and (3) showing that the DNA of leukemic cells contains sequences homologous to those of DNA synthesized with the reserve transcriptase using Rauscher sarcoma RNA as a template.

Similar findings were made in human breast carcinoma in which evidence for a "temperate" cycle for the mouse mammary tumor virus was demonstrated. Spiegelman's studies prove that some of the molecular events described in the life cycle of the virus in rodents exist in humans, but they do not irrefutably establish that these events are in man associated with carcinogenesis, although one gains the feeling that the weight of the evidence favors a viral etiology of human leukemia and breast cancer.

Of particular interest is the epidemiology of Burkitt's lymphoma. Burkitt described a form of lymphoma that occurs in children and has a peculiar propensity to invade certain organs. In tropical Africa Burkitt's lymphoma is the most common tumor in children and is found in the kidney, adrenal and salivary glands, liver, testicles, and other sites. A special feature is the frequent involvement of the jaw. Although Burkitt's tumors have been described even outside Africa, the incidence of Burkitts's lymphoma seems to be high only in limited areas. The disease has not been described in the northern part of Africa, and few cases have been reported in South Africa. Clusters of high frequency have been found in countries surrounding Lake Victoria, and geographic pathologists have concluded that a high-altitude barrier delineates the zone in which the disease is endemic.

Closer examination of the climate in which Burkitt's tumor flourishes has suggested an influence of rainfall (mean minimum of 20 inches) and of temperature (mean minimum of 60° F). Altitudes below 5000 feet, high rainfalls, and high temperatures are favorable breeding grounds for a variety of arthropods. These conditions are believed to favor the infection of some insects by a virus which, when transmitted to man, causes the disease. Thus, the distribution of Burkitt's tumor in Africa resembles that of sleeping sickness, and the similitudes between the epidemiologies of Burkitt's tumor and sleeping sickness lead to the conclusion that an insect must transmit both diseases.

Investigators in several laboratories have found a herpes-like agent in Burkitt cells cultured *in vitro*. But the role of the herpes-like virus in the pathogenesis of the disease is far from clear. It is not known whether it is a passenger or the causative agent.

Apparently, two different types of herpesviruses exist. One causes lip lesions, and the other cervical lesions; the former has antigenic properties different from those of the latter. Thus, herpesvirus type 1 infects only above the waist, and herpesvirus type 2 causes infections below the waist. Type 1 is most likely transmitted by kissing, whereas type 2 is transmitted through sexual intercourse. Therefore, herpesvirus has been called the virus of love. The herpesvirus, which is icosahedral and measures 100–150 nm in diameter, is a DNA virus containing double-stranded DNA. Its capsid is composed of 162 cylindrical capsomeres, each made of five or six rodlike structural units.

Herpesvirus is a nuclear virus. In fact, contact of the virus with the cytoplasm destroys its activity. It is not known how the virus escapes passing through the cytoplasm during its release from the cells. Whether it is encapsulated in a vacuole for transport from nucleus to cell membrane or whether it travels through preconstructed tunnels has not been determined.

The virus genome dictates the biosynthesis of its own protein. However, it uses the cytoplasmic machinery of the host to elaborate the proteins. The viral protein is made within the cytoplasm and transported, again by some unknown mechanism, to the nucleus where it complexes with viral DNA.

The two major forms of herpesvirus infections are the productive form and the so-called nonproductive form. In productive infections, fully infectious particles are made that ultimately kill the cells. In nonproductive infections, no complete viral infectious particles are assembled, but the viral genome is undoubtedly present in some form within the nucleus because the cells that do not contain herpetic viral particles can, under appropriate conditions, produce infective virus.

Like most viruses, the herpesvirus alters the properties of the cell membrane through: (1) changing permeability, (2) promoting cell fusion, and (3) preventing contact inhibition. Little is known of the molecular changes responsible for these alterations of the cell membrane.

The primary infection with herpes occurs in childhood. Usually the victim is unaware of the infection because it does not differ from ordinary childhood infections.

Herpes simplex is a benign disease that leads to an epithelial inflammatory process on the lips (cold sores) or the cornea. Occasionally, it may affect the central nervous system, causing severe encephalitis. Herpetic infections of the cornea result in keratitis, which was a serious condition before optical therapy with IUDR and BUDR was available.

When Epstein and his collaborators succeeded in culturing Burkitt cells from a Burkitt's lymphoma biopsy, they observed a virus referred to as the Epstein-Barr virus. The virus found in Burkitt's tumors has been partially purified, and detailed electron microscopic studies revealed it to be a member of the herpes group. The capsid is made of 162 capsomeres, has a diameter of 900 Å, and is coated with a distinctive layer of an amorphous substance (300 Å thick).

Werner and Gertrude Henle prepared an immunofluorescent antibody to the virus and demonstrated its presence in healthy people as well as in patients with Burkitt's tumors. The incidence of positive serum samples ranged from 80–90% in all persons tested. Since only a small number of individuals develop lymphoma, it must be concluded that if the EB (Epstein-Barr) virus is responsible for Burkitt's tumors, its transforming activity must in some way be aided by other factors.

The correlation between the presence of antibodies to type 2 herpesvirus and cervical cancer is impressive. Among 35 women with cervical cancer, 81% had evidence of exposure to type 2 herpesvirus. In contrast, among 58 women without cancer, only 25% possessed antibodies to the virus. Such results have been reported by two laboratories, that of Nahmias and that of Melnick. Even more impressive are findings indicating that the antigens were present in the preinvasive and invasive stages in patients with cervical cancer. The antigen was restricted to the cancer and was not found in surrounding normal tissues of cancer patients nor in normal individuals. Thus, the appearance of the antigen is an early event preceding the transformation of a normal cell into a neoplastic one.

11

AGING

INTRODUCTION

There have been many legends and many jokes about the aged. The old have been respected and despised, canonized and sometimes left to die. Philosophers, scientists, and even explorers have sought to protect men from growing old.

The fear of aging stems partially from the unpredictability of old age, which Shakespeare expressed well in *The Tempest* (act 4, scene 1, line 191): "And as with age his body uglier grows so his mind cankers." But in *Much Ado About Nothing* (act 1, scene 1, line 14), Shakespeare wrote: "He hath borne himself beyond the promise of his age."

What is age? It is a part of life and the prelude to an inescapable death. We celebrate our birthdays to remind ourselves of our age. Age is the duration of our existence measured in years, or, more specifically, in units of time.

Such a definition hardly encompasses all aspects of age; for example, some people become older sooner than others, as the two quotations of Shakespeare indicate. Moreover, what is time? I suppose that the answer to the latter question is most important, but it is not within the realm of knowledge of the physician to answer this puzzling question. The Bible promises that the age of humans will be three score and ten, but it never promises eternity, at least within the shelter of earthly bodies.

In ancient Rome the average life span was 22 years, and it is now 74 years in the United States. Longevity has increased not because aging has been prevented or postponed, but because we have reduced infant mortality, secured better nutrition, eliminated most devastating infections, and prevented and cured some cancers and heart conditions. In fact, if in the near future we were to find a cure for cancer, atherosclerosis, and other diseases associated with age, it is not likely that the average life span would increase much. Of course the cure would certainly benefit some individuals. Even if we were able to prolong the life span, there is at present no way to prevent the deleterious consequences of the physiological aging process.

Life starts after conception when the ovum is fertilized by the spermatocyte, who among his 500,000 peers released in a single ejaculation succeeded in reaching the coveted prize. Through fecundation, ovum and spermatocyte blend into a single cell that divides. The early embryo attaches itself to the walls of the uterus and within this marvelous environment—which provides rich food, heat, and shelter from noise and trauma—the embryo develops into a fetus. No human consciously recollects the shock of its transfer from the mother's womb to the cruel environment of this earth. Certainly the newborn cannot comprehend the need for the rude slap on his behind, the touching of all parts of his body by masked people, the uncomfortable feeling of being bathed, and the horrible taste of the nipple of the bottle and its artificial contents that are destined to nuture him. Hopefully the human subconscious does not remember these momentous events either.

The infant grows, learns to walk, builds muscles and bones, and his brain develops. At puberty sex hormones are secreted, pubic hair grows, girls' breasts develop, and boys learn to like girls and vice versa. After the age of 20, humans may continue to develop their muscles, but they usually stop growing; and around 40, if not earlier, the hair becomes gray and in men falls out. Wrinkles appear and glasses may be needed for reading because of inability to accommodate. From that time on there are two possible courses: one either becomes sick or grows into a ripe old age free of obvious disease. Today atherosclerosis, cancer, and autoimmune diseases are among the major causes of death in the elderly. In those who survive without disease, the hair becomes white as snow, the skin is covered with wrinkles like an ocean stirred by a gentle wind, the voice weakens but the mind may remain agile, and the gait may be stable and sturdy.

MANIFESTATIONS OF AGING

The Introduction of this chapter gave a temporal description of aging. Biologically, aging is a progressive departure from ideal organismal performance, which in most cases places the aging individual at a physical disadvantage over the adult. Aging is a physiological process that nobody escapes.

A loss of accommodation is part of the process of aging, but a diabetic cataract is a consequence of disease. Although we will briefly discuss the mechanisms of aging, our knowledge of aging is mainly at the descriptive stage.

Among the major manifestations of aging are a decline in ability to perform muscular work, a loss of sexual prowess, and irreversible (except by cosmetic surgery) alterations of the anatomic features of the face and other parts of the body. A study of men and women 75 years old has established the frequency and the types of losses of function that occur with age. Functional losses affect the brain, nerves, heart, lungs, endocrine system, and immunological functions. Some of these changes are summarized in Table 11-1.

Such changes in function could result from either a decline in the ability of each individual cell to perform its function, or a decrease in the total number of body cells. It is often believed that the latter alternative prevails, but available evidence is by no means compelling. Certainly there is cell loss in the aged, but whether such loss is responsible for the changes observed in aging is debatable. Indeed, one may lose large portions of many organs without showing signs of aging (liver, kidney, and even brain).

The autopsy of a person who quietly dies of old age is boringly unimpressive. The viscera are smaller, the muscles are atrophied, and part of the cell population in parenchyma and muscle is replaced by connective tissue.

We know nothing for certain about what causes aging. There have been as many theories about aging as there are investigators studying aging (there are, however, few of them), plus a few theories that emerged in the brains of armchair philosophers.

POSSIBLE CELLULAR MECHANISMS IN AGING

We have seen already that the mammalian body contains three types of cells with respect to their life cycle: (1) some that divide

TABLE 11-1. Functional losses accompanying aging

Type of loss	Remaining function (%)*
Nervous system	
Brain weight	56
Memory	?
Speed of response	?
Brain blood flow	80
Number of nerve fibers	63
Nerve conduction velocity	90
Number of taste buds	36
General metabolism	
Body weight (in ♀ only)	88
Body water content	82
Basal metabolic rate	84
Maximum work rate for short burst	40
Hand grip	55
Heart	
Cardiac output at rest	70
Kidney	
Number of glomeruli	56
Glomerular filtration rate	69
Kidney plasma flow	50
Lung	
Maximum oxygen uptake during exercise	40
Maximum ventilation volume during exercise	53
Voluntary maximum breathing capacity	43
Vital capacity	56
Endocrine	
Adrenal activity	?
Gonadal activity	?
Immune system	
Immunodeficiency both humoral and cellular	

* Figures are the approximate percentages of remaining function in the average 75-year-old man, taking the average value found in the 30-year-old man as 100% (from Nathan W. Shock, *Scientific American* 206:100–110, 1962).

constantly, e.g., those of the skin and the epithelium of the gastrointestinal tract; (2) those that divide only after stimulation (the liver cell after partial hepatectomy or other kinds of liver damage); (3) those that never divide (neurons).

It would appear that with aging there is not only a decrease in the number of cells that do not divide, but there is a slowing down of the replacement of rapidly growing cells. Wound healing and repair of fractures are slow and often impaired in the aged. The livers of old rats regenerate much slower than those of young rats.

It is conceivable that the loss of cells with highly specific functions, such as the cells responsible for immunological reactions or certain neurons, might be critical to the aging process.

One's fear of age largely stems from the damage that it does to the brain. Some may fear impotence and infertility, but most men and women are resigned to accept the restrictions that age imposes on athletic prowess. However, no one accepts mental deterioration and senile dementia. Although many volumes have been written on brain damage caused by aging, there is no established correlation among the anatomical, histological, histochemical, and biochemical events, and the clinicopathological changes.

The difficulty in relating cellular changes observed in organs of old individuals to the cause of aging is best illustrated by a brief description of the changes that take place in the brain with aging. These changes include: (1) loss of cells, (2) reduction of extracellular space, (3) appearance of lipofuscin granules in the nerve cells, (4) development of granulovascular changes in the cells, (5) appearance of neurofibrillar structures inside the cells, and (6) formation of senile plaques.

If it can be said that aging is generally associated with loss of neurons, it is impossible to define clearly the relationship between age and neuron loss. The extraneural pool shrinks in the aging brain. It has been suggested that such anatomical changes may be associated with a decreased capacity for transport of ions and small molecules. Intracellular deposits of yellow, green, or brown fluorescing granular pigments called lipofuscin are found in the cells of aging brains. The number of such granules increases with age. The origin of the granules is not clear, but lysosomal enzymes have been found in association with these granules. The functional relationship between these granules and aging is not known. The granulovascular changes are characterized by the presence of small vacuoles in the neuronal cytoplasm or at the base of dendrites of degenerated neurons.

Senile plaques are composed of a core amyloid surrounded by masses of degenerated dendrites and axons. Their association with senility is not clear. They are absent in brains of some patients with senile dementia but have been found in brains of individuals who until dead were intellectually and emotionally intact. In some brain neurons intracytoplasmic tubules are arranged in compact parallel bundles that push the nucleus to one side as they fill the cytoplasm. The molecular origin and the significance of these structures remain unknown.

Alteration in the cells responsible for immunological defenses may be of particular significance in aging. The aged respond immunologically differently from the young: there is a decrease in the immunological defenses and an increase in the inability to distinguish self from nonself. Both cellular and humoral types of immunity decline markedly with age. This causes immunodeficiency, which can be traced to an age-related decline in the ability of the lymphocyte to proliferate when stimulated with antigens. The decrease in defense mechanisms obviously renders the aged more susceptible to infection and possibly to cancer.

The increase in autoimmune responses could cause various diseases suspected, but certainly not proven, to be of an autoimmune nature (e.g., atherosclerosis), and it may be responsible for the cellular loss observed in the aged, particularly in the brain.

Aging is, however, not restricted to cells; it also involves extracellular material such as lens protein, elastic tissue—especially that in the arteries—and the collagen framework of the body. Lens epithelial cells are progressively altered to form fibers that contain special proteins transparent to light. The older fibers, which are at the center of the lens, may be replaced by new fibers emanating from the epithelial surface. With age there is progressive reduction in the rate of replacement of older fibers. The older fibers become rigid and dehydrated, and accommodation is impaired. Elastin and collagen are special fibrous proteins. As for all proteins, their properties are determined by their special amino acid sequence. The fibers are linked together by smaller molecules forming cross-links. The incidence of cross-links increases with age and reduces the elasticity of both elastin and collagen. Changes in the properties of elastin are probably responsible in part for the increased rigidity of the vasculature. Changes in collagen may cause wrinkles.

BIOLOGICAL CAUSES OF AGING

Thus, aging is associated with decrease in cell function, loss of cells, alteration of intracellular components, and increased incidence of some diseases. These manifestations raise a number of questions. Which changes are critical to aging, and, more importantly, which are responsible for death? Are the changes programmed or caused by environmental changes? Are the changes caused by damage to a single or to several molecules?

Programmed Aging

Programming of aging could take place by two mechanisms: programming of the genome, as is the case for the red cell and the epithelial cells of the skin, or intramolecular programming of individual molecules. (Programming of the genome is discussed in Chap. 5.)

The notion that aging can be caused by programming of the genome is largely based on studies of the senescence of classes of mammalian cells grown in culture. Cultured human fibroblasts go through three phases: a primary culture, a period of vigorous growth, and a progressive decline. Similar findings were made with lung tissue and chick fibroblasts. Yet when embryonic tissues were used they were much easier to culture. When a clone of bone marrow cells is transferred to mice in which the bone marrow has been depleted by radiation, it grows well after the first four transplants, but it dies with further transplantations. All these findings suggest that the life span of cells in culture is limited. Whether the findings made *in vitro* can be related to those observed *in vivo* remains to be seen.

Maintenance of the optimal activity of an enzyme such as β-galactosidase resides in its conformation. Maintenance of the conformation requires energy. It is conceivable that this energy is progressively lost, contributing to the loss of activity of β-galactosidase. How much such an event could contribute to loss of cellular function is not known, but it does take place *in vitro*.

Heating the enzyme glucose-6-phosphate dehydrogenase, an enzyme essential to the hexose monophosphate shunt, decays the enzyme progressively. When the enzyme is heated, it is much more thermolabile when it is prepared from actively growing cultures than when it is obtained from cells on the decline.

There is no appropriate molecular explanation for the programming of cellular aging. A number of hypotheses, none of which is established, have been proposed. They include (1) the loss of one or more substances needed for the maintenance of cellular integrity (there is no evidence that such substances exist); (2) the unwinding of a molecular clock which at the end of its course unleashes a sequence of suicidal steps in the cell; and (3) programmed errors or restriction of transcription or translation.

Somatic Mutations and Aging

One of the most appealing theories of aging is the somatic mutation hypothesis. The postulate is that as cells age gene expression is permanently altered. The somatic mutation theory is not new, but it has been newly interpreted. We have proposed that aging may result from a somatic mutation that may result from restrictions to repair or from error-prone repair. Burnet has extended these views by proposing that the cell loss observed in aging could also result from the repeated assault of injuries to one or more molecular constituents. These injuries lead to the changes in gene expression that take place with age. The mutations are believed to alter the conformation of the enzymes involved in DNA repair and replication. The changes in the enzyme structure lead to error-prone DNA synthesis, which in turn results in somatic mutations.

There is, however, little evidence indicating that the somatic mutation of aging is genetically determined. Environmental factors might be the principal agents in causing somatic mutations. As mentioned, UV light, X-rays, cosmic rays, and many chemicals inhaled or ingested (in food and drink or iatrogenically) can damage DNA; part of the damage may be repaired, part may never be repaired. Although a number of authors have reported a decrease in DNA repair synthesis in various tissues of older animals, the exact role of DNA repair in aging remains to be evaluated.

Certainly as we grow older all parts of the body are exposed to greater amounts of physical agents, such as ultraviolet light, cosmic rays, γ- and X-rays, and various chemicals in food and the atmosphere.

We have already considered the damage that such agents may cause to DNA and the consequences that follow if the DNA is not repaired. However, damage to body constituents may not be restricted to DNA. It may also affect cytoplasmic constituents (e.g., mitochondria, endoplasmic reticulum), the machinery involved in protein synthesis (messenger RNA, tRNA, and ribosomes), the components of the cell membrane, or even extracellular products such as cartilage and connective tissue.

Another question raised by the aging of cells and their progressive loss of function and viability is whether environmental injury that causes such changes afflicts one or more molecules in the cell. For example, if somatic mutations are the primary cause of aging, it would seem that DNA must be the primary target. Yet it is also conceivable that the injuries alter the relationship between DNA and proteins. Such changes could modulate gene expression, and there is evidence that they could be transferred from one cell generation to another.

Many of the environmental agents that may cause mutation affect molecules other than DNA, and at present it is not possible to evaluate how much the combination of molecular injury to DNA and other macromolecules such as RNA, proteins, and lipoproteins may contribute to the process. It would seem, however, that two types of proteins could be particularly sensitive to injury: proteins that are scarce and those with very low turnover. For example, only a few molecules of some of the enzymes regulating the rate of various biochemical pathways may be present in the cell, and if they are not replaced in time after injury irreversible damage may occur. Proteins with slow turnover, such as histones, when damaged might remain altered for long periods of time if they are not replaced, as may well be the case in the brain cell. Damage to brain cell histones may permanently distort gene expression in those cells. Similarly, the slow turnover of collagen and elastin may explain why damages caused by free radicals produced by X-rays, cosmic rays, or chemicals may accumulate in these tissues.

Immunologic Theory of Aging

Aging could be caused by injuries to the cells of all tissues or to a specific type of cell. If the latter is true, damage to the cells responsible for immunologic defenses is believed to be the most likely cause of aging. Impairment of both the humoral and the cellular immune defenses could cause many if not all the changes observed in the aged.

CONCLUSION

Aging is inescapable and its causes are unknown. Nevertheless, it may some day be within our power to prolong life and physical and mental fitness. If this ever happens, each individual and society may have to learn to cope with these new blessings.

BIBLIOGRAPHY

Barnett, L. (1968): The Universe and Dr. Einstein, Bantam Books, New York.
Bernard, C. (1966): Introduction a L'Etude de la Medecine Experimentale, Garnier-Flammarion, Paris.
Bertram, J.S., and Heidelberger, C. (1974): Cell cycle dependency of oncogenic transformation induced by N-methyl-N'-Nitronitrosoquanidine in culture, Cancer Res. 34:526–537.
Bichat, X. (1973): Recherches Physiologiques sur la Vie et la Mort (Menetrier, J., presentation) edited by Gerard & Co., Verviers, Marabout Universite, Belgium.
Bonner, J. (1965): The Molecular Biology of Development, Oxford University Press, New York.
Burnet, F.M. (1973): A genetic interpretation of aging, Lancet (Sept. 1) 480–483.
Burnet, M. (1966): Natural History of Infectious Disease, Cambridge University Press, London.
Busse, E. W. (ed) (1973): Theory and Therapeutics of Aging, Medcom Press, New York.
Duncan, L.C. (1931): Medical men in the American Revolution, Medical Field Service School, Carlisle Barracks, Pennsylvania.
Edelman, G.M. (1976): Surface modulation in cell recognition and cell growth, Science 192:218–226.
Einstein, A. and Infeld, L. (1938): The Evolution of Physics (The

growth of ideas from early concepts to relativity and quanta), Simon & Schuster, New York.

Finean, J.B. (1961): Chemical Ultrastructure in Living Tissues (Kugelmass, I.N., ed.), Charles C Thomas Publisher, Illinois.

Gamow, G. (1966): Thirty Years That Shook Physics (The story of quantum theory), Doubleday & Company, Inc., New York.

Goldstein, S. (1971): The biology of aging, N. Engl. J. Med. 285: 1120–1129.

Good, R.A. and Fisher, D.W. (1971): Immunobiology (Good, R.A. and Fisher, D.W., eds), Sinauer Associates, Inc., Stamford, Conn.

Hochstrasser, R.M. (1964): Behavior of Electrons in Atoms (Structure, spectra, and photochemistry of atoms), (Johnsen, R., ed) W.A. Benjamin, Inc., New York.

Immunodeficiency of Old Age: A Selective Phenomenon? 57th Annual Meeting of the Federation of American Societies for Experimental Biology in Atlantic City, reprinted in Hospital Practice, July 1973, pp. 21–30.

Jaffe, B. (1957): Crucibles: The Story of Chemistry, Fawcett Publication, Greenwich, Conn.

Joly, R. (1964): Hippocrate (Medecine Grecque), Gallimard, France.

Kashgarian, M., Hayslett, J. and Spargo, B.H. (1974): Renal Disease (Brinkhous, K.M., Carter, J.R., and Kinney, T.D., eds.), Upjohn Company, Kalamazoo, Michigan.

Kenny, M. (1973): Pathoparasitology (Thomas, B.A., ed.) Upjohn Company, Kalamazoo, Michigan.

Kitaigorodskiy, A.I. (1967): Order and Disorder in the World of Atoms (Chomet, S., ed.), Springer-Verlag, New York.

Kohn, R.R. (1973): Aging, The Upjohn Company, Kalamazoo, Michigan.

Krumbhaar, E.B. (1962): Clio Medica/Pathology, edited by E.B. Krumbhaar, Hafner Publishing Company, New York.

Lathan, M.C., McGandy, R.B., McCann, M.B., and Stare, F.J. (1970): Scope Manual on Nutrition, Upjohn Company, Kalamazoo, Michigan.

Leboucq, G. (1941): Andre Vesale, edited by Anc. Etabliss. J. Lebegue & Co. Societe Cooperative, Bruselles.

Long, E.R. (1965): A History of Pathology, Dover Publications, New York.

Macleod, A.G. (1973): Aspects of acute inflammation (Thomas, B.A., ed.), Upjohn Company, Kalamazoo, Michigan.

Mishra, N.K. and Mayorca, G.D. (1974): In vitro malignant transformation of cells by chemical carcinogens, Biochim. Biophys. Acta 355:205–219.

Nicolson, G.L. and Poste, G. (1976): The cancer cell: dynamic aspects and modifications in cell-surface organization (Part I), N. Engl. J. Med. 295:197–203.

Nicolson, G.L. and Poste, G. (1976): The cancer cell: dynamic aspects and modifications in cell-surface organization (Part II), N. Engl. J. Med. 295:253–258.

Nicolson, G.L. (1976): Transmembrane control of the receptors on normal and tumor cells (I. Cytoplasmic influence over cell surface components) Biochim. Biophys. Acta 457:57–108.

Orgel, L.E. (1973): Aging of clones of mammalian cells, Nature 243:441–445.

Patterson, R. (ed) (1972): Modern concepts in clinical allergy, Medcom Press, Inc., New York.

Patterson, W.B. (1959): Wound Healing and Tissue Repair (Patterson, W.B., ed.), University of Chicago Press, Chicago.

Pauling, L. (1960): The Nature of the Chemical Bond and the Structure of Molecules and Crystals: An Introduction to Modern Structural Chemistry. (3d ed.) Cornell University Press, Ithaca, New York.

Peterson, A.R., Bertram, J.S., and Heidelberger, C. (1974): Cell cycle dependency of DNA damage and repair in transformable mouse fibroblasts treated with N-methyl-N'-nitro-N-nitroguanidine, Cancer Res. 34:1600–1607.

Pizzarello, D.J. and Witcofski, R.L. (1967): Basic Radiation Biology, Lea & Febiger, Philadelphia.

Rene Vallery-Radot (1900): La Vie de Pasteur, edited by E. Flammarion, Paris.

Rogers, W.P. (1962): The Nature of Parasitism, Academic Press, New York.

Shock, N.W. (1962): The physiology of aging, Scientific Amer. 206:100–110.

Sigerist, H.E. (1958): The Great Doctors, translated by Eden and Cedar Paul, Doubleday & Co., New York.

Terzaghi, M. and Little, J.B. (1975): Repair of potentially lethal radiation damage in mammalian cells is associated with enhancement of malignant transformation, Nature 253:548–549.

Thacher, J. (1967): American Medical Biography, Vol. 2, Da Capo Press, New York.

Thompson, D. (1942): On growth and form (Bonner, J.T., ed.) Cambridge University Press, England.

Trump, B.G. and Arstila, A.U. (eds.) (1975): Pathobiology of cell membranes, Vol. 1, Academic Press.

Van Lancker, J.L. (1970): Hydrolases and Cellular Death, in Metabolic Conjugation and Metabolic Hydrolysis, pp. 355–418, Academic Press, New York.

Van Lancker, J.L. (1975): Hydrolases and cellular death, in Pathogenesis and Mechanisms of Liver Cell Necrosis (Keppler, D., ed.), pp. 25–35, Medical and Technical Publishing Co., Ltd.
Van Lancker, J.L. (1976): Molecular and Cellular Mechanisms in Disease, Springer-Verlag, Heidelberg.
Van Lancker, J.L. (in press): DNA Injuries, Their Repair and Carcinogenesis, Current Topics in Pathology, Springer-Verlag, Heidelberg.
Venzmer, G. (1968): Five Thousand Years of Medicine, translated by M. Koenig, Taplinger Publishing Co., New York.
Virchow, R. (1958): Disease, Life and Man, translated by L.J. Rather, Collier Books, New York.
Walker, K. (1962): Histoire de la Medecine, edited by Gerard & Co., Verviers, Marabout Universite, Belgium.
Wischnitzer, S. (1962): Introduction to Electron Microscopy, Pergamon Press, New York.
Zamenhof, S. (1959): The Chemistry of Heredity (Kugelmass, I.N., ed.), Charles C Thomas, Springfield, Illinois.
Zinsser, H. (1960): Rats, Lice and History, Bantam Books, New York.

INDEX

A

Acetylaminofluorene, and DNA distortion, 136
Acetylcholine, effect on membrane permeability, 239
Acid hydrolases, following cell damage, 258–259
Actinomyces, 97
Actinomycosis, granulomatous reaction, 99
Addison, work of, 12
Addison's disease, 120, 203
Adrenal gland, cortex and medulla, 195
Aflatoxin, and cancer incidence, 86
Aging, 288–296
  autoimmune responses in, 293
  biological causes, 293
  and extracellular material, 293
  functional losses in, 291
  immunological defenses in, 293
  immunological theory of, 296
  manifestations of, 290
  possible cellular mechanisms in, 290–293
  programmed, 294
  and somatic mutations, 294–296
Alchemy, approaches of, 18–19, 22
Aldosterone, in sodium balance, 125

Alkylating agents
  carcinogenesis of, 154–155
  in chemotherapy, 155, 156–157
  effects of, 152–155
  molecular binding, 152–154
Allgemeines Krankenhaus, 13
Amino acids
  atoms and properties of, 46–47
  basic skeleton for, 46
  codons for, 48
  in protein synthesis, 44
Aminopterin, folic acid derivative, 146–147
Amoeboid movement, 211
Ancylostoma duodenale, life cycle, 105
Anesthetics, effect on membrane cytoskeleton, 246–247
Angina pectoris, 265, 268
Angiotensin, and aldosterone, 125
Antibodies
  formation from B lymphocytes, 73
  humoral and cellular, 69–70
  molecular forms, 69
Antibody-receptor complex, and membrane cap, 246
Anticodon, 48, 49

Antidiuretic hormone, in fluid balance, 124
Antigen-antibody reaction, schematic representation, 74
Antigens
  binding to antibodies, 63–64
  in cancer transformation, 249–250
Antimetabolites
  and DNA synthesis, 146–155
  inhibition of protein synthesis, 158–159
Ants
  injuries caused by, 113
  life history, 112
Arterial narrowing, results of, 264
Arthropods
  disease causing, 106, 110–118
  diseases transmitted by, 115–117
Atherosclerosis, 264–273
  blood cholesterol in, 272–273
  cerebral consequences, 268–272
  definition of, 264
  lipid accumulation in, 193
  pathenogenesis of, 264–265
Atherosclerotic profile, 272–273
Atoms
  chemical affinity of, 168
  structure of, 167–170
ATP
  and chemical energy, 49–50
  formation of, 50
Autocatalytic sequence, and rapidity of reaction, 61–62
Autoimmune defects, self-destruction in, 203–204

B

B cells
  and humoral antibodies, 70
  origin of, 70, 72
Bacilli, grouping of, 88
Bacteria
  identifiable structures of, 88–90
  means of classification, 88
  pathogenic, list of, 90–92
  types of, 89
Balantidium coli, 100
Base sequence, of mRNA, 44
Bases
  arrangement in DNA, 45
  purine and pyrimidine, 44–45
Beriberi, thiamine deficiency in, 119, 185
Bernard, Claude, contributions of, 28–29
Bichat, contributions of, 8–9

Biological agents, and disease, 86–88
Biosyntheses, in cytoplasm, 49
Blastema, theory of, 14, 16
Blastomycosis, lesions of, 98–99
Blood cells, schematic representation, 66
Blood circulation
  schematic representation, 58
  steps in, 57
Blood coagulation, 64–66
  fibrinogen splitting in, 63
  steps in, 65
Boerhave, contributions of, 7–8
Boils, and inflammation, 66–67
Brain infarct, extensiveness of, 270
Brain softening, description of stages in, 270
Breast cancer
  events in, 273–274
  possible virus etiology, 284
Broussais, views of, 10–11
Body fluids
  contents of, 123
  and electrolyte imbalances, 121–125
Body water, distribution of, 123
Burkitt's lymphoma, and Epstein-Barr virus, 284–286

C

Callus, resorption and replacement, 77
Cancer, 273–287
  agents causing, 140–145
  chemicals causing, 140–142
  clinical pathology of, 273–274
  of the liver, 279
  of the lung, 274
  metastases of, 143
  pathogenesis of, 275–276
  survival advantages, 140
Cancer cell,
  definition of, 274
Cancer cells, in vitro transformation, 247–250
Cancer studies, difficulties in humans, 282
Cancers and viruses in humans, 283–287
Candida albicans, description of, 97, 98
Capsids
  cylinders of, 41
  of viruses, 94
Carbon tetrachloride, effect on protein synthesis, 160
Carcinoembryonic antigens, importance of, 24*
Carcinogenesis
  testing for, 281–283
  two stages, 275–276

INDEX

Carcinogens
  and alteration of gene expression, 144
  examples of, 280–281
  in humans, 277–283
  metabolism of, 142
  profession related, 277–278
  in vitro transformations, 276–277
Catabolic pathways, and autodigestion, 262
Catalysts, characteristics of, 174–175
Catalytic proteins, role of, 131
  composition of, 180
  injuries to, 167–194
  pathology of, 182–190
Cell adhesion, 212–214
  binding in, 213
  determining forces, 212–213
  irreversibility of, 214
Cell communications, 214–216
  calcium in, 215
Cell differentiation, and molecular interaction, 34
Cell injuries, primary, secondary, and critical, 252–253
Cell membrane, *see also* Plasma membrane
  alteration in cancer cells 247–250
  biosynthesis of, 224–226
  communication network, 51
  enzyme interference, 239–240
  difficulties in studying, 218–219
  erythrocyte alterations, 235–237
  functional injuries, 239–250
  functions of, 40, 208–217
  inner cytoskeleton disruption, 244–247
  lipid bilayer, 51
  lipid-protein relation, 222
  liquid crystal form, 223
  molecular organization of, 219–224
  pathology of, 235–250
  receptor injuries, 242–244
  role of endoskeleton, 52
  schematic representation, 52
  secretion interference, 240–242
  selective permeability of, 51
  structural injuries, 235–239
  structure of, 218–219
  transport interference, 240
Cell movement, mechanisms of, 212
Cell structure, 40–53
Cells
  effect of molecular composition, 34
  ions in, 123–124
  molecular degradation, 35
  molecular organization, 40–41

  schematic representation, 43
  shapes of, 41
  specific function of, 33–34
Cellular death
  catabolic enzymes in, 257–259
  correlation of events in, 259–262
  importance of, 252
  molecular alterations in, 258–259
  reflections on, 251–262
Cellular destruction, embryonic, 82–83
Cellular metabolism, regulation of, 216–217
Cellular necrosis, in chronic inflammation, 68–69
Cellular organelles, of macromolecular complexes, 40
Cellular pathology, 13–17
Cellular proliferation, triggering of, 75
Cerebral hemorrhage
  hemodynamic factors in, 272
  symptoms of, 272
  theories of mechanism of, 271
Cerebrosides, composition of, 193
Cervical cancer, and herpesvirus antibodies, 286–287
Chalones, proliferation inhibitors, 75, 78
Chediak-Higashi syndrome, microtubule injury in, 247
Chemical bonds
  covalent, 171
  double, 172
  hybrid, 172
  ionic, 170
Chemical messengers, in intercellular communications, 54–55
Chemical pathology
  origins of, 17–32
  in 19th century, 28–32
Chemical reactions
  rate of, 173
  reversible, 173–174
Chemoreceptors, of inflammatory cells, 243
Chemotactism, inhibitors of, 243–244
Chloramphenicol, effect on protein synthesis, 158, 159
Cholera, electrolyte imbalance in, 129
Chromosomal anomalies, examples of, 82
Chromosomes
  composition and function, 40
  morphology of, 42–43
  number of, 35
  sex, 35
  structure of, 37–38
Chromatin, schematic representation, 36

Chronic inflammation
  character of, 67–68
  events in, 68–69
Cigarettes, as carcinogens, 278–279
Circle of Willis, arteriosclerosis in, 269
Circulatory system
  purpose of, 57
  structure of vessel walls, 56
  vessels of, 56
Clonorchis sinensis, life cycle, 103–104, 106
Coat protein, self-assembly in TMV, 42
Cocarcinogens, role of, 141
Cocci, characteristics, 88
Coccidioidomycosis, disease reactions in, 99
Codon, bases sequence, 48
Coenzymes, defects in, 184–187
Complement, components of, 70
Concanavalin A, effect on membrane cytoskeleton, 246
Congenital anomalies, causes of, 82–84
Congenital spherocytosis, sodium permeability in, 236
Congestive heart failure, fluid imbalance in, 125, 126
Contact inhibition, specificity of, 212
Cooper, contributions of, 12
Coronary artery occlusion, consequences of, 265–266
Coronary heart disease, 265–268
Corticosteroids, effects of binding, 200
Corvisart, contributions of, 9, 13
Covalent interactions, in cell regulation, 216
Cretinism, and thyroxine metabolism, 120
Cryptococcosis, agent of, 98
Cyclic AMP
  effects of, 200
  in cancer transformation, 249
Cyclic GMP, in cancer transformation, 249
Cycloheximide, in translation interference, 158, 159
Cystathioninuria, and vitamin B_6, 185
Cytocavity network, 226–234
Cytochrome b_5, properties of, 222
Cytochrome oxidase, effect of cyanide, 189
Cytoplasm, composition of, 43
Cytoplasmic degradation, hypothetical mechanism, 261
Cytoplasmic metabolites, flow of, 260
Cytosine arabinoside
  in leukemia, 152
  properties of, 152
Cytosol, 43, 49

D

Dehydration, and electrolyte imbalance, 125
Defense
  autocatalytic and nonautocatalytic, 62
  first line of, 60
  overlapping of reactions in, 62
Defense mechanisms, 59–79
  defective, 125–127
Desmosomes
  in cell joining, 53, 215
  damage in skin cancer, 238
Detoxification, by ER enzymes, 49
Diabetes
  causes of, 201–203
  symptoms and effects, 202
Diethylstilbestrol, in cattle feed, 86
Differentiation, in gene expression, 36
Dipalmitoylphosphatidylcholine
  biosynthesis of, 241
  surfactant agent, 241
Diphtheria toxin, in protein synthesis, 160
Diphyllobothrium latum, 103
Disease
  definition, 1, 80
  and substrate alteration, 190–194
DNA
  alteration of base sequence, 139
  alterations in hereditary diseases, 81
  bacterial, 90
  base pairing in, 46
  consequences of injuries to, 137–138
  damage to, 163–164
  in mitochondria, 51
  primary injury to, 132–133
  repair mechanism, 60
  structure of, 44–45
  structural gene of, 44
  Watson-Crick model, 47
DNA breaks
  and chromosome alterations, 133
  effect of alkylating agents, 132–133
  effect of X-rays, 132–133
  major kinds of, 132
DNA repair, 133–137
DNA synthesis, interference with, 145–155
DNA virus, characteristics, 94
Double-strand breaks
  and cancer induction, 137
  after X-irradiation of DNA, 137
Drugs, side effects, 86

E

Edema, major causes of, 124–125
Egypt
  disease records, 3
  early disease concepts, 4
  early embalming, 3
Electrolyte imbalance, causes of, 126
Electron dense lines, of cell membrane, 53–54
Electron transport chain
  integration of, 50
  molecules in, 40
  and oxidative phosphorylation, 41
Electrons, properties of, 169–170
Elephantiasis, and filaria, 110
Encephalomalacia, results of, 270–271
Endocrine glands, 195–196
  absence of, 201
  feedback inhibition, 200, 205, 206
  hyperplasia of, 206
Endonuclease
  in DNA repair, 133–134
  effect on X-radiated DNA, 136
Endoplasmic reticulum
  effect of barbiturates, 227
  functions of, 44
  membranes of, 43
  and protein synthesis, 159–160
  smooth and rough, 43–44, 226
Energy
  efficiency of aerobic pathway, 50
  sources in cytoplasm, 49
Entamoeba histolytica, 100, 101
Environment, conflict with, 84–118
Epithelia, effects of disruption, 60
Epithelial cells, migration and proliferation, 75
Epitheloid cells, in granulomatous reactions, 69
Enzyme activation, means of, 181–182
Enzyme activity, modulation of, 62
Enzyme block, of DNA synthesis, 146
Enzyme defects, protein synthesis alteration, 138–139
Enzyme inhibition, and disease states, 253
Enzyme inhibitors
  competitive and noncompetitive, 178–179
  properties of, 179
  reversible and irreversible, 178
Enzyme molecules, absence of, 183
Enzyme reactions, 175–180
  active center, 175
  initial velocity, 177–178, 180
  velocity constants, 176
  enzyme-substrate complex, 175–177

Enzymes
  activators of, 179–180
  allosteric site, 181
  associated with cell membranes, 219
  in cell membrane, 52–53
  defective, 183
  defective regulation of, 184
  effect of pH, 180
  effect of temperature, 180
  groups of, 180
  increased activity, 184
  life cycle, 180–182
  life span, 182
  in liver regeneration, 78
  in polymorphonuclear granules, 182
  specificity of, 181
  in zymogen granules, 181–182
ER, see Endoplasmic reticulum
Ergotism, effects of, 85–86
Erlich
  and chemotherapy, 32
  work of, 31–32
Erythrocyte damage, in type of hemolytic anemia, 237
Erythrocyte membrane, proteins in, 236
Erythrocytes
  causes of drug induced hemolysis, 236
  conversion of precursor, 38
  effects of glycophorin and spectrin, 244
  immune lysis, 235
  lysis by detergents, 235
  membrane damage in aging, 236–237
  osmotic shock, 235
  potassium loss in hemolysis, 235
  prelytic membrane defects, 235
  programmed death in, 254
  steps in membrane lysis, 236
Ethionine, in protein synthesis, 160
Eukaryotic cells, 88
Existing knowledge, development of, 33–58
Excision repair, of UV-irradiated DNA, 134

F

Facilitated diffusion
  nonregulated, 209
  regulated, 210
Fibrin, formation of, 65
Fibroblasts, movement of, 211–212
Fibrosis, in chronic inflammation, 69
Fluorinated analogs, resistance to, 151–152
Fluoro-derivatives, activity of, 151

5-Fluorouracil
  properties of, 150
  tumor effect, 150–151
Focal cytoplasmic degradation, concept of, 260
Folic acid
  derivatives of, 146–147
  formula, 148
Folic acid antagonist
  in cancer therapy, 147
  formula, 149
Fracture repair, steps in, 77
Fractures, 75–77
  schematic representation, 76
Fungi
  incidence of, 96
  major categories, 98
  major classes of, 97
  structure of, 97

G

Galactosemia, phosphoglucomutase inhibition, 189–190
Galen, theories of, 5
Galileo, and the compound microscope, 6
Gammaglobulins, absorption by pinocytosis, 211
Gangliosides, in cancer transformation, 248
Gap junctions
  in cell communication, 216
  damage in hepatomas, 238
Gastric ulcer, schematic representation, 61
Gaucher's disease, cerebroside accumulation, 192–193
Gene alteration, in inborn metabolic errors, 81
Gene expression, 33–40
  in bacteria, 81–82
  in human chromosome, 82
  and programmed death, 38–39
  in reaction to injury, 39
Genome repression, of red cells, 38
Giardia intestinalis, 100
Glauber, discoveries of, 22
Glomerulonephritis, immune system in, 129
Glucose, cell membrane transport, 209
Glucose-6-phosphate dehydrogenase, in hemolytic anemia, 183
Glucose transport, energy requirement, 210
Glucosuria, in diabetes, 202
Glutamine analogs, in DNA synthesis, 147–148
Glutamine antagonist, formula, 153
Glycophorin
  membrane protein, 52, 218
  structure of, 218

Glycoproteins, in cell metabolism regulation, 216
Glycosides, effect on membrane enzymes, 239
Glycolysis, 49–50
Goiters, iodine deficiency in, 204
Golgi apparatus, role in cancer transformation, 248
Gout, disease process, 184
Gramicidin A, and ion transport, 224
Granulation tissue
  formation of, 75
  in fractures, 76–77
Granulomatous reaction
  character of, 69
  in tuberculosis, 69
Growth hormones, hyperplasia of, 120

H

Harvey, contributions of, 6
Health, definition, 1, 80
Heart, vasculature of, 266
Heart rupture, factors in, 267
Helix, of DNA, 45–46
Hemoglobin
  function, 188
  and iron deficiency, 188–189
  molecular structure, 188
Hemophilia, chromosomal defect in, 127
Hepatomas, and cholesterol synthesis, 184
Hereditary defect, in a-antitrypsin deficiency, 126–127
Hereditary diseases, 81–82
  germ cell injury in, 81
Herpes simplex, 286
Herpesvirus
  cell membrane alterations, 286
  properties of, 285–286
Hippocrates, writings of, 4
Histamine
  activation of, 63
  release in inflammation, 66
Histones, basic proteins, 38
Histoplasmosis, description, 99
Hodgkin, work of, 12
Hormonal action, schematic representation, 1
Hormonal agents, and cancer cells, 120–121
Hormonal diseases, 201–206
Hormone degradation, 201
Hormone imbalance, 120–121, 195–206
Hormone level
  causes of decreases in, 120
  increases in, 120–121

Hormone receptors, defects of, 205
Hormone secretion, regulation of, 198–200
Hormone synthesis, defects in, 204
Hormone transport, 197–198
  defects in, 204
Hormones
  action site specificity, 198
  cytoplasmic receptors, 198
  and intracellular transport, 210
  mode of action, 196–198
  polypeptide and nonpolypeptide, 195
  regulatory defects, 205–206
  second messengers of, 200
Human subacute encephalitis, viral origin, 95
Humoral antibodies, actions of, 70
Humoral theories
  of early Greece, 4
  in 18th and 19th centuries, 7, 9, 14
Hunters, contributions of, 11
Hypercholesterolemia, and receptor deficiency, 242
Hypoparathyroidism, effects of, 201
Hypophysis, lobes of, 196
Hyperglycemia, in diabetes, 202
Hypertrophy, and hyperplasia, 77–79

I

IgG, molecular structure, 71
Immunity, 69–71
Immunodeficiency diseases, and inadequate development, 127
Inducers, in embryonic development, 83
Inflammation, 66–69
  signs of, 67
  Inflammatory reaction, in cellular death, 74–75
Injury, excessive response to, 127–130
Insulin
  activation of, 62–63
  as chemical messenger, 55–56
  regulation of secretion, 199
  structure and synthesis, 196–197
Intercellular integration, 53–57
  microscopic features, 53–54
Intestinal cell, absorption by microvilli, 211
Iniazid, carcinogenicity of, 281, 282
Isotopes, properties of, 170

K

Kidney
  glomerulus of, 205
  role in water balance, 124

Krebs' cycle, 49
Kuru, slow virus disease, 95–96
Kwashiorkor, protein deficiency in, 118, 190

L

Laennec, and the stethoscope, 10
Laplace, work of, 27
Lavoisier, contributions of, 25–27
Leeuwenhoek, microscopic studies, 6, 87–88
Leishmania, infection sites, 101
Leishmania donovani, 101
LETS, absence in cancer cells, 248
Leukemia, virus etiology of, 284
Lice, human infestation, 113–114
Life, origin of, 121–122
Lipid bilayer, membrane model, 222–224
Lipidoses, accumulating compounds, 191–192
Lipofuscin granules, in aging, 292
Liver cirrhosis
  cellular reactions, 193
  histological distortion in, 79
  sodium retention in, 125
Liver regeneration
  biochemistry of, 78
  and DNA synthesis, 39
  of specific cells, 78
Loa loa, 105, 111
Lymphocytes
  and antibody-forming cells, 70
  function of, 57
  in granulomatous reaction, 69
Lysosomes, 227–228
  in cellular death, 258
  labilizers of, 228
  stabilizers of, 228

M

Macromolecular complexes, examples in cells, 40
Macromolecules, role in the cell, 40
Malaria, transmission of, 101–102
Malpighi, histological studies, 6
Median cerebral artery
  effects of occlusion, 270
  cerebral hemorrhage in, 272
  thrombosis of, 268
Medical schools
  of 18th and 19th centuries, 7–12
  of Vienna, 12–13
Medicine, role of, 1
Medieval times, medical knowledge in, 5

Meiosis, steps in, 34
Membrane microfilaments, structure and function, 244
Membrane microtubules, structure and function, 244
6-Mercaptopurine, effects of, 148
Mersalyl, inhibition of adenyl catalase, 239–240
Mesosomes, of bacterial membranes, 90
Messenger RNA (mRNA), in protein synthesis, 44
Metabolism, inborn errors of, 81
Metals
  chelation of, 63
  deficiency of, 188–189
Metastases, 274
Methotrexate, folic acid derivative, 146, 147
Microfilaria, life cycle of, 104–106
Microorganisms, and disease, 87–88
Mitochondria
  and aerobic pathways, 50–51
  appearance of, 51, 229
  composition of membranes, 231
  fatty acid oxidation in, 50
  function of cristae, 230
  function and structure, 229–232
  morphological alterations, 234
  origin of, 232–234
  osmotic behavior, 230
  and protein biosynthesis, 232–233
  relations of membranes of, 229–231
Mitochondrial DNA
  coding properties of, 233
  properties of, 232–233
Mitochondrial proteins, endoplasmic reticulum synthesis of, 233
Morgagni, studies in organ pathology, 7, 9
Mushrooms, toxin of, 85, 244–245
Mutations
  inherited 138–139
  of proliferating genes, 139–140
  somatic, 139–145
  types of, 160
Myasthenia gravis, neuromuscular junction anomalies, 239
Mycoses, superficial and deep, 86
Myelin, importance of, 55
Myocardial infarct
  clinical observations, 267–268
  healing and aneurysms after, 267
  histology of, 267
  pathological appearance, 266
Myocardial ischemia, effect on cell membranes, 237

N

Neuron
  fibrillar extensions, 55
  schematic representation, 56
Nervous system
  functions of, 54
  schematic representation, 55
Nitrosamines, carcinogenetic properties, 279
Noncovalent interactions, in cell regulation, 216
Nonhistones, acidic proteins, 38
Nucleotides, analogs in cancer therapy, 148–152
Nucleus
  functions of, 42
  structures in, 42
Nutritional deficiencies, 118–119

O

Oncogene theory, and cancer cell transformation, 145
Oncogenic viruses, identification of, 283
Oocyte, schematic representation, 35
Operon, activation of, 37
Operon theory, of gene expression, 36
Orbitals
  of the carbon atom, 172
  electronic, 168–170
  hybrid, 172
  molecular, 169
Organisms, four categories, 96
Oxidative phosphorylation, events in, 50
Oxygen, discovery of, 23–28

P

Pancreas
  glands of, 196
  structure, 202
Paracelsus, work of, 19–21
Parasites
  and host relationship, 114, 118
  of man, 87
Parathormone, and calcium regulation, 206
Parathyroids, hyperplasia in, 121
Paroxysmal cold hemoglobinuria, hemolysis in, 236
Pasteur
  and disease microorganisms, 29–31
  fermentation studies, 29–31
  work with tartaric acid, 29

Pathogenic mechanism, general, 161
Pathology
    foundation of, 5–12
    purpose of, 2
Peptide bonds, of polypeptides, 47
Pernicious anemia, and vitamin B_{12}, 119
Peroxisomes, 228–229
Phagocytosis
    cell membrane effects, 246
    events in, 67
    of necrotic tissue, 74
Phalloidines
    effects on cell membrane, 245–246
    and liver necrosis, 245
Pheochromocytomas, and adrenal hyperplasia, 120
Phenylketonuria
    characteristics of, 183
    metabolic disorder, 81
Pi bonding, in ethylene molecule, 172, 173
Placenta, and birth defects, 83
Plasma membrane
    appearance of, 41
    carrier-mediated transport, 209
    as cell barrier, 208
    in cell respiration, 217
    in genome expression, 217
    hormone receptors, 210
    selective permeability of, 209
    transport regulation, 208–211
Plasmodia
    life cycle, 101–102
    pathogenic types, 101
Plasmodium falciparum, schematic representation, 102
Platelets
    in blood coagulation, 65
    character of, 57
Pneumonia
    inflammation in, 67
    schematic representation, 68
Polyene antibiotics, cell membrane damage, 238
Polymorphonuclears
    function of, 57
    in inflammation, 66–67
    passage of, 67
    types of, 66
Polynucleotide, composition of, 44–45
Polynucleotide chains, of DNA, 45
Polypeptide chain, character of, 47
Porphyria hepatica, porphyrin levels in, 184
Posterior cerebral artery, occlusion of, 269
Postreplication repair, of DNA, 135
Potassium, effect on protein synthesis, 217

Potassium ions, concentrations across membranes, 52–53
Prehistoric times
    diseases present in, 2
    knowledge of medicine, 2
    records of, 2
Priestly, findings of, 24–26
Programmed death, 253–254
    environmental effects, 254
    examples of, 254–255
Proinsulin, covalent structure, 197
Prokaryotic cells, 88
Properties, of coagulation, inflammation, immunity, 61
Prosthetic groups, and protein activation, 62
Protease, activation of, 63
Protein activation
    mechanisms of, 62–64
    by molecular splitting, 63
Protein coat, of TMV, 40
Protein inactivation, mechanisms of, 63
Protein synthesis, 44–49
    events in, 44
    in rough ER, 44
    schematic representation, 45
Proteins
    amphipatic, 52
    in circulatory fluids, 57
    functions in cell membrane, 52–53
    relation to lipid bilayer, 51–52
    roles in cells, 40
Protozoa
    movement of, 100
    properties of, 100
    structure of, 99–100
Provoked death, 255–262
    ATP-regenerating capacity in, 256–257
    molecular mechanisms in, 256
    point of no return, 256
Pus, contents of, 67
Pyrimidine antagonists, formula, 153

R

Radiation, effects of, 85
Receptors
    for catecholamines, 242–243
    in various tissues, 243
    types for histamine, 243
Red cells, function of, 57
Renaissance
    chemistry and medicine of, 19–23
    medical advances in, 5–6

Restoration, of lost tissues, 72–81
Reverse transcriptase, of leukemic cells, 284
Rhodopsin, and transmembrane channels, 222
Ribosomes
   composition of, 43
   on endoplasmic reticulum, 43–44
   linkage interference, 159–160
   role in protein synthesis, 44
Ringworm, characteristics of, 97
RNA, structure of, 48–49
RNA polymerase, in transcription, 155
RNA synthesis, injury in, 164
RNA virus, characteristics, 94
Rokitansky, work of, 12–14
Rutherford, work of, 23–24

S

Scabies, mite infection in, 112
Schwann, theory of, 14–16
Scurvy, capillary walls in, 186
Senile plaques, 292
Sickle cell anemia
   and metabolic distortion, 81
   hemoglobin alteration in, 138–139
Simmonds' cachexia
   and anterior hypophysis, 120
   tropin defect in, 203
Skeletal muscle, glucose penetration, 210
Sodium
   combined with chloride atom, 169
   electronic orbitals of, 168
   reabsorption mechanisms, 124
Sodium ions, concentrations across membranes, 52–53
Sodium pump, and respiration levels, 217
Sodium retention, causes of, 125
Specificity units, injury to, 131–166
Spleen infarct, schematic representation, 128
Spores, bacterial, 90
Sporotrichosis, effects of, 99
Stroke
   by arterial obstruction, 268–269
   by hemorrhage, 268
Structural proteins, cell components, 131
Substrate accumulation, 191–193
Substrate depletion, of DNA synthesis, 146
Substrate inhibition, types of, 179
Subunit theory, of membrane structure, 221
Sugar, deoxypentose in DNA, 45
Surfactants
   breakdown of, 241
   causes of deficiencies, 242
   importance in lung diseases, 241
Sylvius, work of, 22–23

T

T cells
   and cellular antibodies, 70
   origin of, 70, 72
Taenia solium, life cycle, 103
Template unit, composition of, 132
Tetralogy of Fallot, anomalies in, 83–84
Thalassemia, and breakdown of mRNA, 158
Thalidomide, effects in pregnancy, 82
Thioguanine, properties of, 149–50
Thrombi
   formation of, 65
   in vessel occlusion, 65–66, 269
Thrombus formation, defense overreaction in, 128
Thymine dimers, in UV radiation of DNA, 135–136
Thyroid hormone, steps in regulation of, 199–200
Tight junctions
   of cell membranes, 53–54
   function of, 54, 215–216
Toxins
   accidental intake, 85–86
   effect on neurotransmitters, 239
Transfer RNA
   character of, 49
   in polypeptide chain, 49
Translation
   events in, 158
   injury in, 165
   pathology of, 158–161
Transcription, injuries to, 155, 158
Trauma
   examples of, 84–85
   results from, 84–85
Trepanation, incidence in prehistoric times, 3
Trichinella spiralis, and trichinosis, 103
Trichomonas vaginalis, 100, 101
Trypanosomes, and sleeping sickness, 100–101
Tubercle
   formation of, 69
   schematic representation, 70
Tumors, and somatic mutations, 140
Typhoid fever, lymphoblast proliferation in, 129

U

Urine, formation of, 205
Unit membrane, theory of membrane structure, 220–221
UV irradiation, and cancer, 140

V

Valinomycin, effect on membrane permeability, 224
Van Helmont, contributions of, 22
Vasculature
 of the brain, 268, 269
 of heart, 266
Vasodilation, in wound healing, 73
Venom
 of honeybee, 111–112
 of wasp, 112
Vesalius, contribution of, 6
Viral DNA, repair of, 137
Virchow, contributions of, 14–17
Viruses
 and cancer in animals, 93
 and cancer in humans, 93
 as cancer cause, 142
 and cell transformation, 142–143
 effect on host DNA, 143
 molecular composition, 93–94
 pathogenic, list of, 94–95
 size of, 93
Vitamin A, deficiency of, 187
Vitamin C, and hydroxyproline, 186
Vitamin D
 and calcium absorption, 187
 production of, 186
Vitamin deficiency
 causes of, 119
 effects of, 119
Von Gierk's disease, glycogen accumulation, 193–194

W

Water and sodium balance, 124–125
Worm infection, 102–106
 intermediate host in, 103
 from undercooked food, 103–104
Worms, pathological, 107–109
Wound healing, 73–75
 example of, 73
 factors in retardation of, 75

X

Xeroderma pigmentosum, cancer incidence in, 144
X-irradiation, and cancer, 140

SCHEELE MEMORIAL LIBRARY

3 6655 00028245 0
RB112 .V37
Van Lancker, Julien/Molecules,

RB   Van Lancker, Julien
112   L. 1924-
.V37
          Molecules, cells,
          and disease

DATE			
APR 5 1994			
MAY 1 6 1997			

CONCORDIA COLLEGE LIBRARY
BRONXVILLE, N.Y. 10708

© THE BAKER & TAYLOR CO.